# OUTWITTING
# CRITTERS

SOME OTHER BOOKS BY BILL ADLER, JR.

*Outwitting Squirrels*
*Outwitting the Neighbors*
*Outwitting Toddlers*
*The Expert's Guide to Backyard Birdfeeding*
*Impeccable Birdfeeding*
*Baby-English: A Dictionary for Interpreting the Secret Language
of Infants*
*Tell Me a Fairy Tale*

# OUTWITTING CRITTERS

A Surefire Manual
for Confronting Devious Animals
and Winning

# BILL ADLER, JR.

Illustrations by John L. Heinly

THE LYONS PRESS

*To Richard, Joanna, Florence, and David*

Printed in the United States of America

Designed by Alma Hochhauser Orenstein

10  9  8  7  6  5  4  3

Library of Congress Cataloging-in-Publication Data

Adler, Bill, 1956–
    Outwitting critters : a surefire manual for confronting devious animals and winning / Bill Adler, Jr. ; illustrations by John L. Heinly.
        p.      cm.
    Originally published: New York, NY : HarperPerennial, c1992.
    Includes bibliographical references.
    ISBN 1-55821-523-9 (pbk.)
    1. Pests—United States—Handbooks, manuals, etc.   2. Wildlife pests—United States—Handbooks, manuals, etc.   3. Pests—Control—United States—Handbooks, manuals, etc.   4. Wildlife pests—Control—United States—Handbooks, manuals, etc.   5. Pests—Handbooks, manuals, etc.   6. Wildlife pests—Handbooks, manuals, etc.   7. Pests—Control—Handbooks, manuals, etc.   8. Wildlife pests—Control—Handbooks, manuals, etc.   I. Title.
    [SB605.U5A65    1997]
    648'.7—dc20                                          96-27500
                                                              CIP

# CONTENTS
## A Cast of Animals

# ACKNOWLEDGMENTS

The acknowledgments is the scariest part of a book to write. While misspelling a word or misplacing a comma is a permissible sin (not that I'm implying that I've done that), forgetting to thank somebody who helped is not forgivable. So, hoping that I haven't left anybody out, let me start by thanking Beth Pratt-Dewey, Adler & Robin Books' senior editor. Her capable research skills and keen knowledge of the outdoors contributed so much to this book.

HarperCollins has a fantastic editor: Without Cynthia Barrett's guidance and patience I could never have completed this book—or even started it. After working with Cynthia on two books, I can't wait to write another with her.

Carol Dana came to the rescue when I needed her.

I am always happy to thank Jane Dystel, my agent in these affairs, for helping me through the many stages of this project.

Thanks also to my wife, Peggy, for not minding my absence too much while I pounded away at the keyboard; and thanks also to Peggy for taking time from writing her most recent book to give advice on mine.

A cast of experts and victims of critter unrest contributed to this book as well. My thanks to: The Great Plains Agricultural Council

Wildlife Resources Committee, the National Coalition Against the Misuse of Pesticides, the Cooperative Extension service at the University of Nebraska–Lincoln, John Adcock, John Anderson, Keith Aune, Laurie Bingaman, Richard Brenner, Donald Cochran, Bob Davidson, Mark Fenton, Frank Godwin, John Hadidian, Niles Kinerk, Larry Manger, Susan Grace Moore, Carl Olson, Wayne Pacelle, Peggy Pachal, Tom Quinn, Lowell Robertson, Cal Saulnier, Becca Schad, Dave Sileck, Sherri Tippie, Betsy Webb, Mark Westall, Heidi Youmans, and Bill Zeigler.

# INTRODUCTION

Every writer needs a kick in the pants to get going. A wallop sometimes. In my case, I had dawdled long, and kept saying I'd start writing tomorrow. We all do it. Just ask any editor.

But inspiration jolted me one evening into beginning to write this book. As I was resting in bed I heard this crumbly sound inside the bedroom wall. It went on for about five minutes and I was convinced that what I had heard was plaster tumbling in our ninety-year-old house.

Convinced, that is, until I saw a brown and black furry object scurry across the floor. A chipmunk!

For the next hour I slinked around the house, a flashlight in one hand to let me peer into darkened crevices, and a camera flash in the other to temporarily stun the little critter. Now never mind the fact that I had no way of actually capturing the chipmunk (my wife had suggested a nylon laundry bag, but I was certain that wasn't going to work), I confidently felt that I would figure out a way as soon as the two of us confronted each other.

The day after I saw the chipmunk I went to the hardware store and bought a Havahart trap, baited it with roasted peanuts and acorns, set it on the most sensitive setting, and waited. After a few days, nothing had happened. And I found my memory had grown fainter and my imagination bolder: Was that really a chipmunk and not a rat? What if there's a whole family of chipmunks (or worse!) breeding in the walls of our house?

I wish that trap would spring. After all, my credibility as an out-witter of urban wildlife expert is on the line. I'll keep you posted.

In the meantime, this incident does serve to remind us that critters are everywhere. From a distance they are cute and interesting to watch. At other times, they are not so cute.

Here's a short list of animals that are cute, and the circumstances in which they are not so cute:

### Deer

CUTE: Anytime, as long as you don't have a garden.
NOT CUTE: When you have a garden; or when you wonder about Lyme disease.

### Roaches

CUTE: When your one-year-old catches one in his hand, and your video of the event wins the top prize on *America's Funniest Home Videos*.
NOT CUTE: Every other time.

### Squirrels

CUTE: When in somebody else's yard.
NOT CUTE: At your birdfeeder or in your attic.

### Bears

CUTE: In the zoo.
NOT CUTE: In your tent.

### Gophers

CUTE: In the cartoons.
NOT CUTE: On the golf course.

### Bats

CUTE: From a distance.
NOT CUTE: In your attic.

### Skunks

CUTE: When de-scented.
NOT CUTE: When they've just met your dog.

### Alligators

CUTE: In an animal park.
NOT CUTE: In your swimming pool.

### Pigeons

CUTE: On postcards of Venice.
NOT CUTE: Throughout the rest of the known universe.

Animal invasions are increasing. As the suburbs expand, animals have to go somewhere.

Obviously, some retreat, following their shrinking habitat. But others try to stake a claim in the same places that people live. Indeed, houses can become desirable for animals (as well as people), especially if they are situated in formerly undeveloped areas that still retain streams, trees, and other vegetation.

Houses can offer suitable protection for certain nocturnal animals and an excellent food supply. Raccoons, skunks, and opossums are especially interested in discovering the benefits of the suburban life. Garbage cans, barbecue remnants, and pet food left outdoors are manna to a skunk or raccoon. Vegetable gardens and orchards are favorite desserts for deer.

Most of the time animal invasions are just annoying. You know what I mean: A deer munching on your strawberries; a squirrel dining on the seed you intended for the birds; a colony of ants that has taken up residence near the pile of Cheerios your ten month old built beneath her high chair.

Annoyance is a perfectly good reason to try and outwit critters:

There's no reason to feel guilt or shame about this. No one disputes the fact that the woods are a bear's home, but the woods are a big place, so why does the bear have to occupy your eight square feet? The raccoons were in the neighborhood first—that's for sure—but by leaving out garbage, pet food, and other "edibles," humans have given raccoons every reason to stay in the suburbs and no reason to move back to the woods.

Sometimes, however, animal invasions are more than just annoying: They can be dangerous. Coyotes and alligators have a fondness for small dogs, for example. Roaches and ants carry disease; so do mice. Deer may carry ticks bearing Lyme disease. Rabid raccoons are proliferating in the East.

There is no way to tell by looking at an animal whether it has rabies, except that rabid animals often exhibit no fear of humans. For example, nocturnal animals, such as skunks and raccoons, may approach people in a friendly manner during the daytime, but then bite when touched or petted. As a precaution, children—and adults—should stay away from any wild animals, but particularly those that appear unafraid or sick.

Pets should be kept away from wild animals as well, at least to the degree possible. Vaccinations exist to protect dogs and cats from rabies. Since cats often roam more freely than dogs, and enjoy hunting other animals, cats do face some risk of being bitten by a rabid animal. Along with contracting the disease itself, a cat can pass it along to its owner.

Anytime it is bitten by another animal, your pet should be brought to a veterinarian. This applies even to a dog who has been vaccinated. And, of course, a person bitten by a wild or domesticated animal should immediately see a doctor.

Thousands of species come into contact with humans. And, under the right circumstances, almost any critter can become annoying or even dangerous. However, because it isn't possible within these pages to talk about *every* critter that bothers people, I had to decide which animals to exclude.

Take cows for instance. My wife, Peggy Robin, was driving across California when traffic was held up because a herd of cows was crossing the road. Somehow one cow became separated from the pack and wandered up a cliff-like path above the road. The cow started mooing, in a deep, frightened voice. "I knew what was going to happen," Peggy said. "The cow was going to jump." And it did, missing Peggy's car by inches. Boom.

That's one good cow story. I looked for more, however, and discovered that people aren't usually bothered by cows. So they're one of the animals I decided I could safely exclude.

Other species, however, clearly passed the annoyance test and are covered in the pages that follow. There, you'll find plenty of specific suggestions for how to shoo, ambush, discourage, deter, or otherwise outwit these unwelcome guests.

But there are also some general rules of thumb that apply no matter what creature is in question:

1. *Avoid pesticides and poisons.* You may kill the creatures you want to get rid of (or at least some of them), but you will also kill other animals you don't want to get rid of. Animals that are resistant to the poisons—or able to outsmart the poisons—will be back in greater quantity and strength.

2. *Use territoriality to your advantage.* Most animals are territorial: They establish a certain range and tend to stay there. That's a disadvantage because it means an animal will want to stick where it is no matter how much you shout at it. However, once you coax the animal to another territory—either by moving it or by convincing it to move on its own accord (by removing the animal's source of food or water, for example), it will tend to stay away. This brings us to rule number three.

3. *Remove sources of food.* This means don't leave the pet food outside, cover the vegetable garden with netting, bolt the garbage can lid down. Most of the time, an animal has nothing to do except to eat, or try to eat. If there's nothing to eat where you are, the animal will go someplace else. This rule applies equally for outdoor critters like raccoons and indoor ones like mice. Rule number three sounds very elementary, and it is; but it is also very effective.

Rule number three also has a corollary, the motto of all zoos:

3A. *Don't feed the animals.* When you give a wild animal a handout it will expect more. Feed the squirrels and you will have friends for life (or at least for a year, which is the squirrel's average life expectancy).

**4.** *Don't handle a dangerous animal yourself.* Animals bite. That's because biting is one of their major defenses. And the consequences of being bitten go well beyond the agony of the immediate wound.

**5.** *Block the animal's entrances.* If there are any holes between the inside of your house and the outside, put mesh or brick or some other substance over it. For the purposes of outwitting critters, holes include heat exhaust pipes, chimneys, and pet doors. Unfortunately, deactivating a pet door to keep a raccoon out, also keeps your cat in. This brings us to Rule number 5-A: *Some solutions create more problems.*

**6.** *Erect barriers.* Fences, netting, Plexiglas covers, even minimoats all can make it difficult or impossible for an animal to get where it wants to go.

**7.** *Think like the critter you want to outwit.* Get down on your hands and knees and look at the world from that animal's point of view.

**8.** *Keep your house in good condition.* This is different from the recommendation that you seal potential entrance holes. The more your house resembles a wild place—like the woods—the more attractive your house will be to critters. Painting surfaces, trimming bushes, installing exterior lights, mowing grass, repairing leaking porch roofs—all of these maintenance measures help to create a distinction between your dwelling and theirs.

Well, enough of this introduction. On with the show.

# ALLIGATORS AND SNAKES

## ALLIGATORS

The North has cold, snowy, icy winters. Many people regard that as a drawback.

But the South, the Gulf states in particular, has alligators. And not just in zoos and adventure parks.

For example: A boy was leisurely riding his moped along a road in Travis County, Texas, when his path was blocked by a four-and-a-half-foot alligator. Not your everyday road block, but a mere baby as far as these creatures go. The police were summoned, and the two deputies who arrived hog-tied the alligator, who managed to escape the ropes twice, before the deputies got the knots in place on the third try. The alligator was transported to a nearby lake, where it was set loose.

Or maybe you'd rather take a swim in Long Beach, North Carolina? Although alligators prefer fresh water, when ponds and lakes become shallow in the summer, gators will seek out water wherever they can. One August in the recent past, an alligator made his way to a heavily-touristed section of the ocean at Long Beach and spent several days playing hide-and-seek with the tourists and police. An officer involved in the hunt said the police were only able to capture the gator after they "slipped a four-wheeler between him and the water and cut him off."

Then there's the story of Adolph in Indianapolis. It seems this four-foot, nine-inch gator escaped from the Old Indiana Amusement Park there and stayed away for seven weeks. Adolph had been freed by teenagers who thought it would be fun to watch this critter crawl around. He eventually showed up at a park lake, where an animal trainer spent a month trying to lure Adolph back with chickens. No luck. They finally had to drain the lake to get him.

As if it's not bad enough that the outdoors is crawling with these things, some people have acquired pet alligators for their homes. They make great watchdogs; their bite is far worse than their bark. But if you ever try to get rid of the alligator, *that's* not so easy. One Colorado resident who advertised his three-and-a-half-foot "Wally" in the newspaper received only one inquiry. When the potential buyer found out that Wally would eventually reach six or seven feet, he slithered out of the deal.

Alligators. On a National Geographic special, they're fascinating; without the television screen for protection, they're frightening. But that's no reason to move north.

Native to Southeast wetlands, the American alligator is our largest reptile and may grow to lengths of eighteen feet. Most probably don't reach that size, and today a large one is around twelve feet long. Yearlings may be eighteen inches long, and they grow about a foot a year for several years, under good conditions. So only a dedicated owner would keep a pet gator for years.

Heavy dorsal scales protect the alligator's back, and the rest of the body is covered with dry scales. Unlike mammals, alligators are cold-blooded, that is, they don't have any internal control to maintain a constant body temperature. They prefer warm weather, hot, even. But not too hot. In fact, alligators cannot survive body temperatures above ninety-five degrees Fahrenheit. Poachers who want a perfect carcass let the sun do their dirty work and leave an immobilized animal in the sun for several hours until it dies. Left on their own, alligators stay on the move in their territories, trying to find somewhere warmer or somewhere cooler.

In fact, it is the alligators' unending search for something better that often brings them to our yards, resorts, and pools—and into conflict with us. The bad news is that they like many of the same habitats that humans do: sunny waterfronts, golf courses, ponds, and canals. Any freshwater, and even some brackish or salt water, is suitable habitat for gators. The good news is that they're territorial, so usually you only have to deal with one at a time.

Alligators are unrepentant carnivores; they have no worries about heart disease and cholesterol. Alligators eat meat, and they're well equipped for their diet. They have sharp, conical teeth and several replacement sets of them at that. But the teeth don't regenerate indefinitely, and an old alligator may be nearly toothless. Unfortunately, it may not be possible to determine the age and condition of the gator's teeth until it's too late. Under the very best of conditions, an alligator may live to be fifty years old.

Alligators eat fish, amphibians, some birds, other reptiles (especially turtles and snakes), carrion, and small mammals. Little particles of blood or flesh in the water often given an alligator the first indication of nearby food. When they sense these appetite stimulants, they'll violently lash their heads from side to side, snapping their jaws, looking for the tasty tidbit. Overlooking the question of mating, for gators most things fall into two categories: food and nonfood. Generally gators regard people as nonfood—unless unwise folks cement a food-human association by feeding them.

Gators are always looking out for the next meal; they're opportunistic feeders and eat whenever given the chance. If they are too full to eat something, they'll secure a meal under a log at the bottom of a lake. Just a snack for later.

Cold weather makes them sluggish, and in the winter, they become inactive and feed less. In some northern parts of their range, alligators go into a sort of hibernation and burrow into the muddy ground to wait out the cold.

With the advent of spring, an alligator's thoughts turn briefly from food. Males and females come together for about three days, for courtship and mating. After they copulate, the male leaves for his own home range, and the female remains in hers to build a nest and lay eggs.

There is little outward differentiation between the sexes, and unless you can get close enough to manually examine the internal genitals—not advised—it's tough to know which is which. But a dead giveaway is an alligator guarding a nest; you can bet that's a female. About two months after mating, she builds a nest, of mud and whatever vegetation is most handy, about three feet high and five feet wide, and she scoops out a cavity where she will deposit thirty to fifty eggs. The nest is usually constructed in a shady area not far from water.

Some of the most dangerous alligators are females protecting their nests from intruders; females have reportedly attacked bears who ventured too close to the nest. On one occasion, a female attacked an airboat of Florida game commission census takers. Even an alligator has trouble harming a boat, but the officers stayed away. Females will attack almost anything that ventures too close to a nest. A nesting female embodies maternal protection as an armored, snapping, fight-to-the-death fury.

Alligator eggs are about three inches long and the same diameter at both ends. The female does not need to incubate the eggs; in most of the alligator's range, the air temperature is warm enough to do the job.

The mother stays in the vicinity of her nest while the eggs incubate. During the day, she often rests on the nest with her throat on top of it. In addition to keeping the nest safe from predators, like raccoons who would eat the eggs, she dampens the structure with water that drips from her body after a swim and keeps the eggs from overheating.

Baby alligators dig out of the eggs and then the nest by themselves between July and September; the timing depends on the weather conditions. The maternal instincts are strong, and the mother stays around the young for a few months to protect the small reptiles from predators, including other alligators. Even so, it is estimated that as many as ninety percent of baby alligators die within nine months. Juveniles who survive their first year in the wild can probably survive into adulthood.

In natural history parlance, there are limiting factors that keep animal populations in check. One limiting factor for the alligator is habitat. Once on the endangered species list, the alligator is on the mend. Hunting is regulated and even the environment is a little more benign. Many naturalists speculate that development of golf courses and canals in Florida has actually increased livable habitat for the alligator in recent years and allowed the population to expand. But a burgeoning human population has simultaneously brought about more interspecies contact than either would prefer.

Florida is pretty much the alligator capital. The Florida Game and Freshwater Fish Commission receives nearly ten thousand calls a year complaining about the presence, or behavior, of an alligator. And about a dozen full-fledged alligator attacks are reported in the state each year. Lieutenant Tom Quinn of the state game commission says that many of these encounters could be avoided. He gives one example: "Three men were driving along when they spotted an alligator by the road. They decided to put it into the trunk, and a man was bitten."

Bill Brownlee of the Texas Parks and Wildlife Commission concurs. "People's curiosity gets them in trouble. A gator may look dead, when it's not." So don't get closer to find out. He recommends that people stay at least thirty feet away from an alligator. Texas gets about a hundred alligator complaints each year, and about fifty gators are destroyed by the commission annually.

And in case you've wondered: There are no alligators living in the subway tunnels of New York City. (The rats ate them.)

Given alligators' large size, sharp teeth, and small brains, it seems that we should be the ones to exercise caution, rather than the other way around. Your best strategy for reducing the risk of a gator encounter will vary depending on where you, and/or the alligators, are.

## In Lakes, Ponds, and Pools

If you're outside a major urban area in gator country, any body of water could harbor alligators, so assume it does. Any alligator that's four feet long or bigger is a threat, and if you can see one or know one lives in the area, then it's not wise to swim there or to send your dog into the water chasing after sticks.

People with water dogs need to be especially cautious. Mark Westall, a naturalist on Sanibel Island in Florida, explains that when people throw a tennis ball out into the lake, the disturbance can act like a feeding stimulus, alerting the alligator that a potential meal is at hand. "The alligator looks at that dog and says, 'Well, that's not quite a raccoon, but it looks a little like a raccoon.'" A Texas wildlife official said dogs were like an ice cream treat for alligators.

Remember that alligators are opportunistic feeders; they will always attempt to seize the prey, because they don't know where the next meal will come from. They can't tell the difference between a

retriever's legs and a raccoon's or a deer's—and, generally, they don't waste time trying.

Of course, this naturally brings up the question of whether they can tell the difference between a deer's legs and your child's. Accounts of alligators grabbing dogs naturally make some parents nervous. "They think, 'Well, if the gator is going to take a dog, which weighs ninety pounds, what about my children?' " says Westall. He maintains that alligators look at dogs and say, "There's a funny-looking raccoon." But alligators look at human beings and say, "Here is a predator, here's a killer who is going to kill me." Westall says an alligator naturally wants to keep a respectful distance.

But all too often the alligator has learned that people mean food—garbage, fish entrails, or even purposely fed snacks. An alligator unfamiliar with people will be ignorant of our food possibilities—but what are you going to do, ask it? "Excuse me, Smiley, but has any human ever happened to feed you?" Don't swim in waters with gators. Don't let your kids do it. Or your dog.

Natural bodies of water may not be the only ones harboring alligators. When it gets really dry, alligators may cast a longing eye at our swimming pools and man-made ponds. When natural water is sparse, alligators head for any swimable source of water. Even ponds as deep as twenty feet are attractive to alligators.

From a gator's perspective, water is water. For an alligator a swimming pool is a perfect place to run and hide when confronted by a human. Alligators do not jump into pools to eat people; they go into pools to escape people. It matters little whether there are swimmers in the pool already; fresh water is a safe place for an alligator. Land is not.

Although alligators will venture into pools for protection, they don't like the chlorine. So, left on their own, alligators will get out of a swimming pool as soon as they feel it is safe to do so. Of course it's easier if there are steps in the pool or some contrivance placed there that helps the animal exit—kind of like putting a board in a basement window well to allow a small mammal to get out. Clear the area and the alligator will leave.

Most of the time.

Mark Westall tells a harrowing tale of trying to get one out of a subdivision's pool. "He sank to the bottom of the pool, so I sank a rope and slipped it around his neck. I pulled him out of the pool and started dragging him across the fairway to let him go in the lake.

"An alligator is like a mule. The more you pull them in one direction, the more they want to go in the opposite direction. They think you're dragging them to their death. Why else would a human being put a rope around their neck? So I'm dragging him with about fifteen feet of rope, and about halfway across the fairway, the rope went limp. Before I could turn around to see what had happened, the alligator was brushing against my legs as he ran by; he was about an eight-footer. If I was bowlegged he would have run right between my legs.

"And the next thing I know, the alligator is leading me to the water. The alligator saw the lake and saw safety. He could have easily grabbed me if he wanted, but he wanted to get to the water. I looked like I had a Doberman on the end of a leash. I loosened the noose, and he did a racing dive into the lake, and that was that."

## On the Move

The water isn't the only place you might encounter a gator. Alligators also like to move around a bit, so you may meet one on its way from one place to another.

Where are the alligators going? "They used to live in a marsh, now it's a golf course; they're going to move into the pond there. Also the other thing is alligators naturally are going to move from pond to pond—especially in the spring when they are trying to mate," explains Westall. And each yearling alligator needs to find a home, too. They will also be on the move after heavy spring rains or during floods.

With today's subdivided world, ponds and blocks of land are geometrically and pleasingly arranged for developers and architects, not for the ease of alligators. In the old days, alligators would go from a low area, walk through a marsh, and get to another low area; today they have to get up and walk across areas that we've filled in to make land for houses. When the alligators want to leave their pond to try to find a mate in the pond across the street, they walk across property.

"If an alligator is sitting in your driveway, he's not there to hunt your kid, he's just trying to get to the pond across the street," Westall explains.

Westall is especially tolerant of the gators and, with some other naturalists, actually initiated a program to provide "escorts" for Sanibel's roaming reptiles. "It's the only place in the entire state that has such a program," he says.

The escorts were once a group of naturalists, which included Westall and some others who were licensed by the state to move alligators around. Today the responsibility is handled by the local police force, trained by Westall and the other gator guardians, although Westall still escorts or moves an occasional alligator. The escorts are there to keep the gator out of trouble, not guide it to any one location.

The wary tolerance of the residents of Sanibel Island helps to keep conflicts at a minimum there. As Westall explains, "On Sanibel we realize the alligators are going to walk around periodically. If you live here for a while, you get used to it." However, he says problems grow as more and more newcomers arrive in the area—people who "freak out when they see an alligator walking down the street."

But for the time being at least, escorts seem to work, and if you have regular alligator crossings in your area, consider implementing this kind of escort plan. However, first check out your plan with your state's fish and wildlife agency, since contact with alligators is strictly regulated. Local naturalists or outdoor clubs are a good place to find help too.

## In the Yard

But what if some Sunday you decide to mow the lawn and there's a big gator parked right in the middle of the yard? You could use the alligator as an excuse to go back inside and watch the ball game. Or, you could go on the offensive.

Actually, that gator should have scrammed for cover as soon as it heard the screen door slam. Clear your throat, yell. Wait a bit. Chances are, it'll leave. Most alligators have a healthy fear of humans; we nearly wiped out the entire species to make a fashion statement, and they haven't forgotten it. However, an alligator on your turf who doesn't flee has lost its fear of humans and is a job for professionals. Call the state wildlife office.

Or maybe the animal is a female with a nest, in which case you should see the mound nearby. Is this a nuisance and danger? Or is this an unparalleled opportunity to observe untamed nature? If you adopt the latter attitude, just bear in mind that you'll lose the use of your yard for about five months. Every year.

Waterfront houses are most likely to have frequent gator visitors. However, if you can keep the animals out of your yard to begin with, you can mow the lawn whenever you want. The key is to make it difficult for the alligator to come from the water up into the yard. Heavy vegetation should work.

People who clear all the vegetation along the waterfront, which a lot of people do because they want to plant sod lawns, actually create an attractive spot for alligators, a regular Coney Island for the green set. In fact, clearing the lakefront is a little like putting out a bird feeder and then getting upset because birds visit. Leave existing vegetation or plant new vegetation along the lake; plants like spartina, cord grass, and leather fern will create a thick, grassy or fernlike buffer along the waterline.

You're no doubt thinking, "Well, the alligator will walk right through those grasses." But the alligator doesn't want to bushwhack through the grasses to get up to the rest of the lawn. It's hard work, and it cuts off a speedy exit for the gator.

Shrubbery prevents alligators from swiftly moving back into the water, their refuge from encroaching humans. As naturalist Mark Westall says, "Remember the old Tarzan movies, when the crocodiles would dive off the bank? Notice how they weren't running through any major stands of vegetation. They were sliding right off the bank into the water." Plant vegetation that creates a buffer between the wild area and your place, and the alligators won't sun on your lawn.

However, this natural solution has some drawbacks. While the vegetative strip might keep gators from sunning on your lawn, it does offer something that alligators find attractive: cover. If the vegetation is located on the edge of your property, that's where the gator will stay. The cover might provide a good place to hunt or even lay eggs. But, if you don't want to go this route, there's another option, according to Mark Westall.

"On Sanibel we let people put up a fence, because a lot of people trust man-made structures more than they do natural systems." If a fence is put in, it should be sunk in the ground a bit to keep alligators from going underneath it. Sanibel's fencing regulations also require people to set the fence six feet back from the waterline to give alligators a sunning spot. "We have a lot of people that want to put a fence in right up against the edge of the water because they don't want the alligator to sun there at all," says Westall. "We think the wildlife has a right to be here also."

In addition to setting fences back from the waterline, Sanibel residents must also set fences in from the side property lines. Residents have to leave a three foot corridor, an alligator alley, so that the alligators—and the turtles and the moorhens—have a corridor they can follow to other habitats nearby.

If you do have an unfenced yard, "Don't leave your dog tied up alone in a yard when there are alligators around," points out one member of the Sanibel Island police force. It seems alligators have a fondness for dog meat. Dogs also aren't very savvy when it comes to alligators and may just invite trouble. They'll bark, run around them, even try to bite them on the nose. All the while the gator is waiting for its chance to chomp.

Cats seem to be a bit smarter. Cats generally steer clear of the reptiles because they don't like the water that much, and despite a reputation for curiosity, cats don't have many questions about alligators. They see an alligator as a potential threat. As a matter of fact, the only time Westall has heard of a cat being taken by an alligator was when one was tied in a yard.

You don't have to live with a nasty reptilian neighbor. States with alligator residents have special programs for nuisance alligators, and they'll send a gator expert to your home to decide how to handle the situation. Often they'll try to relocate smaller animals, but they have to destroy aggressive alligators or larger animals they can't trap.

## In the Dark

Alligator encounters can also occur after dark. In fact, according to Mark Westall, "Alligators do most of their walking at night; they're nocturnal, and they walk under cover of darkness. Just like any good commando, they want to stay out of everybody's way if they can." Luckily, there are some simple steps you can take to help you both keep out of one another's path.

A flashlight is effective for spotting alligators. Like most animals, gators' eyes glow when caught in the light: Adult alligators' eyes glow red; young alligators' eyes glow yellow.

If you take a flashlight, you can spot the alligator at a distance of about fifty feet. Then you can turn around, go back to your house, call the police, and let them escort the gator out of the area. Or, you walk over to the other side of the street and continue on your way. Alligators don't hunt like lions, running things down on land. Alligators are only trying to cross over the land. (That, of course, doesn't mean that it won't try to defend itself if you step on its back.)

If you don't carry a flashlight, the alligator is going to see or hear you coming first, and it's going to sit there and remain calm. It may let you come within six to three feet before it hisses at you. The gator hisses because it's scared, but you're going to have a heart attack. All the more reason to carry a flashlight.

## A Word on Feeding

More likely than not, if you encounter an alligator in your yard or elsewhere, it will either freeze or turn tail and try to get away from you. Frank Godwin says, "Alligators are pretty shy animals, really." He should know. Godwin is president of the Gatorland Zoo, a longtime Orlando attraction. Only occasionally will an alligator aggressively approach a human; these gators think we're a source of food—or even the food.

Problems come in when people try to use food to overcome the gators' natural shyness. A Louisiana country club's experience offers a case in point. The club acquired three alligators for its lake. Every afternoon an audience gathered to watch the golf pro feed the toothy, open mouths. Soon enough the gators were leaping out of the water to gulp down chip shots. What would be next? A caddie? The sad ending to this story is that the animals had to be destroyed. No one wanted them on the course, and it was unsafe to relocate them to the wild once they had begun to associate humans with food.

Lieutenant Tom Quinn, with the Florida game commission, says that alligators that attack humans have lost their fear of us, and that's usually because they've been fed. And whether they're fed marshmallows by tourists or scraps of fish dropped at a fish cleaning station, such alligators will learn to associate people with food. In a gator's mind, people practically are food.

Time and again, wildlife officers discover that the aggressive alligators they have to destroy were once nine-inch-long yearlings fed by some well-intentioned person. The lesson being: Just don't feed alligators of any size or age.

This goes for all wild animals (okay, except birds). It's serious business, says Westall, "So if you see somebody in your neighborhood feeding an alligator, don't say 'Oh, that guy's just taming an alligator,' say, 'That guy may be causing some child to die in this neighborhood.' Feeding the alligators is that type of threat."

The problem with alligators is not with the alligators, but with people. We generally do not like to have large carnivores around where we live. Instead of adapting or moving, we kill the species. We have laws protecting alligators, but it's certain that as more people move to the Southeast, there will be increased pressure on these reptiles.

### ❖ AVOIDING ALLIGATORS

1. Never feed an alligator.
2. When you walk at night through alligator territory carry a flashlight.
3. Create a barrier between your lawn and the alligators' swimming area. This barrier will discourage alligator exploration by making it harder for the alligator to reenter the safety of the water.
4. Don't chain your dog in the yard. (Same thing for your children.)

## SNAKES

A lot of people don't like snakes. I, however, am not one of those people. Not that I think that snakes are exactly cuddly. But most snakes are no bother to humans at all. Sure a snake may curl up with you in your sleeping bag, but most adults can easily escape a snake who shows an interest in getting close.

Statistics help provide some perspective. Of the world's more than two thousand different kinds of snakes, only 250 species are venomous. And, in the United States, there are only 4 poisonous snake species. So, while there are several thousand cases of snake bite every year in the United States, there are relatively few fatalities. Of a thousand reported rattlesnake bites, only three percent were fatal.

Snakes are found throughout the world, with the exception of the cold polar regions. Like all reptiles, they are cold-blooded, unable to internally regulate their body temperature. Instead, Snakes regulate their body temperatures behaviorally, by traveling to cooler or warmer locations as survival demands. Like the alligator, they'll die of heat stroke if immobilized in the sun for long. They live in a variety of habitats from desert to rain forest, treetops to underground. In colder climates, they must hibernate during the winter and, occasionally, snakes of the same or even mixed species will hibernate in the same den, which gives rise, among people with overactive imaginations, to horror stories about nests of rattlers. These nests protect hibernating snakes that will disperse when the weather warms up.

Snakes have no legs but get around quite well using a sinuous motion. Some are even good climbers and prey on birds and eggs. And, amazingly, some even swim.

Snakes' internal organs have evolved in some interesting ways in order to fit inside their long, slim bodies, though the adaptations are more pronounced in some species than others. For instance, most snakes have only one lung, the right one, or have only a vestigial left lung. Internal organs—the kidneys, liver, and reproductive organs—are elongated and staggered in their placement.

Snakes are covered with scales, some of which are specialized. A transparent scale covers the snakes' eyes, in place of eyelids, and these scales are shed each time the snake sheds its skin. Prior to its sloughing off, the skin becomes cloudy, and the reptile has difficulty seeing. They're particularly edgy around this time because they are vulnerable and so they are especially liable to bite. You'll recognize a shedding snake because of its general tattered appearance and cloudy eyes; this is no time to make its acquaintance.

Although they can't hear or see too well (well, they're actually deaf but can sense vibrations), they do have an acute sense of smell. The snake's forked tongue darts out, collects particles, and then is withdrawn and transfers the particles to ducts in the roof of the mouth called the Jacobson's organ. Their tongues and Jacobson's organs function much like our noses and are used to find prey.

Snakes are hunters; some actively seek prey while others merely lurk in a likely spot and wait for prey to arrive. They all swallow their prey whole. They can even swallow prey larger than their heads—the bottom jaw is loosely connected to the skull and can be spread apart. Some snakes have special heat sensors that help them find warm-blooded prey, even in the dark. Their teeth are inclined backward, down the throat, to help move the prey down the digestive tract. Although their teeth are often broken, snakes can grow new ones, and even new fangs. Depending on the size and species, snakes eat a variety of foods: eggs, insects, frogs, crabs, small mammals, or other snakes. Some specialize, but many are generalized feeders.

Smaller prey may be swallowed whole without first being killed. But larger snakes, who eat larger prey, must kill their prey first so it doesn't create such a ruckus going down. They either squeeze the prey to death or use a venom that kills it. Even nonvenomous snakes have exceedingly strong saliva that is designed to help break down complex organic structures—like animals—so even a bite from a nonvenomous snake may be painful. One meal may last for several weeks if the snake isn't very active.

Breeding usually occurs once a year, but because snakes are solitary, finding a mate can be tough. There are no singles bars for snakes. Using their sense of smell, males find receptive mates by following a trail of secretions left by the female of the species. Courtship varies among the snakes from rather tame body rubbing between male and female to ritualistic combat dances between males fighting over a female.

Snakes either lay eggs (they're called oviparous) or carry them within their body until they are ready to hatch and then produce live young (viviparous). Oviparous females lay eggs in clutches and leave them to incubate, although a few species do guard the nest. Their clutches tend to hold between thirty and forty eggs. Snakes that bear live young may have four to fifteen young. Viviparous snakes may live in colder regions than oviparous ones, but the mothers pay the price of being heavier, slower, more vulnerable to predators, and less able hunters.

Snakes are no more fond of humans than most humans are of them. The occasional snake around your garden or yard will make a speedy exit as soon as it knows you are there. But just because the snake hits the road doesn't mean it leaves the neighborhood. Most stick around an area for several months and then move along. They're there because they're busily eating animals that could be pests for you, most likely rodents and insects.

That's why snakes aren't usually a threat to humans. We're simply not the right size to be a meal.

## Poisonous Snakes

There are four poisonous snakes in the United States—the copperhead, rattlesnake, coral snake, and water moccasin. While venomous snakes are found in every state except Alaska, Hawaii, and Maine, it's unlikely that you'll find any of them in your house, with the exception of the copperhead, which is fond of cool, damp places, like basements. Any snake is attracted to the kind of cover usually found around homes.

The copperhead ranges from Massachusetts to northern Florida and west to Illinois and Texas. Although its bite will make you sick, it's unlikely that it will kill you.

Different species of rattlesnake are found throughout the country, from the desert sidewinder to the timber rattlesnake in forests of the Southern and Eastern United States. There are many types of rattlesnakes, and they vary in color. Most are various shades of brown, tan, yellowish, gray, black, chalky white, dull red, and olive green. Many have diamond, chevron, or blotched markings on their back and sides. Like the other pit vipers, they have an elliptical eye pupil and a deep pit on each side of the head midway between the eye and nostril. Learn to identify the rattlesnakes common in your region of the country; don't bother with the others. Of the various types of rattlers, the most deadly is the big diamondback, whose bite may kill fifteen out of every hundred people bitten, according to snake experts.

Coral snakes prefer the warm climes of the south from South Carolina to Florida, along the Gulf states and Texas and west into New Mexico and Arizona. The coral snake is ringed with red, yellow, and black, with the red and yellow rings touching. Other snakes mimic the coral snake's appearance with red and yellow rings, with black rings separating them. Remember: "Red on yellow, kill a fellow; red on black, friend of Jack." Coral snakes are not pit vipers.

The water moccasin lives in ponds, streams, lakes, and swamps from Virginia to Florida, along the Gulf states and into Texas. Like other poisonous snakes, it rarely bites humans. But for safety's sake, you should assume that any swimming snake is dangerous.

Become familiar with local poisonous snakes so you'll recognize one if you see it. Local naturalists, park rangers, wildlife officials, or the staff at local sporting goods stores can often provide you with information on the snakes in your area.

Bill Zeigler, general curator at Miami's Metrozoo, reports that many of the calls he receives about big, dangerous snakes are really about small, harmless ones. There's a big difference between a water moccasin and an indigo snake—a nearly blue-black, docile Florida snake that can reach lengths of six feet.

If you think you've spotted a poisonous snake around your home, report it to the local game or wildlife department immediately. And then keep track of the animal's movements so you can show the official where it went. If you can't point it out, it's going to be pretty tough to find it.

Even though few of us die from venomous snake bites, it's good to know about them. Children are most vulnerable to the bites. Read up on snake bite treatment; in general it's best to get the victim to a hospital as quickly as possible.

A poisonous snake bite causes an almost immediate reaction: swelling, darkening of tissue around the bite, a tingling sensation, and nausea. All snakes have teeth and will leave teeth marks, but only a pit viper will leave fang marks. Wash the wound with soap and water, elevate the wounded area, and keep the victim calm.

A national parks employee in Virginia relates the wrong way to react: "This guy hurt the girl worse than the snake. He was so sure it was a copperhead bite that he crisscrossed the wound with knife incisions and tried to suck the poison out. He had her all cut up, and you could tell it wasn't even a copperhead bite." Leave that stuff to the TV movies.

## Snakes in the Outdoors

If you have snakes around your home and garden, it's probably because you have a good food source for these carnivorous reptiles, or maybe you have a great hiding place: tall grass, accumulated flower pots, damp, dark, cool spaces, heavily mulched flower beds, any garden, brush.

For instance, a large woodpile on the ground attracts field mice. Now snakes are big mouse meat fans and will haunt your woodpile in anticipation of a nice meal. To thwart the snakes, you can either trap and remove the mice or you can raise the woodpile at least fifteen inches above the ground. Animals won't burrow under raised woodpiles, and your snake should curtail its visits.

Don't use poison on the mice or snake. Why not? If you have a snake "problem" it's probably because you have ample food around, like mice. Over the long run, poison will do more harm than good. Mice reproduce at an alarming rate, snakes once a year. If you poison the mice, you'll also poison one step up the food chain, the snakes—and you'll still have mice. But the snakes will take longer to bounce back, and the mouse population will test the limits of the local food supply.

Other likely habitats for mice (or other sources of food for snakes) include high grass, bushes, shrubs, rocks, boards, and junk around the yard. These little rodents also come to feast on spilled birdseed, ill-maintained compost, and Spot's food. What's cover for the rodent is cover for the snake. Snakes love junk. Your old refrigerator out in the yard? A snake mansion. Full of mice and hidey-holes. Get rid of it. By all means cover or enclose your compost heap.

Another snake attractor is a warm spot in the sun. On cool, sunny days, reptiles often sun themselves on rocks and other open spaces, which is why you so often see them smashed on the road. (Or are they just crossing?) Make the sunning spot unattractive; shade it, clean it up, slant it. They'll go elsewhere in search of a place in the sun, somewhere close by for sure, but somewhere you won't have to look at them.

A Laurel, Maryland, woman found a snake sunning itself on her windowsill. She moved from the apartment, but she didn't have to. If you are finding snakes sunning themselves on a windowsill or porch, find a way to discourage the animal. You can install a temporary incline on your windowsill. If you position a piece of wood, metal, or plastic with the incline sloping out- and downwards, the snake won't be able to rest there. But don't be surprised if it shows up somewhere else, although you'll never be sure it's the same snake.

Still, if you're determined to get rid of a particular snake, you'll have to spend some time following it around. If you can close up its dark, secluded hiding spot, the animal will move on to another secure hiding spot. Maybe five feet away. It's hard to anticipate where a snake will want to live—in some old flowerpots, underground, or in your rock garden. Don't get alarmed, but even if you think you don't have snake, you probably do. Small, secretive snakes you never see often live in gardens or wooded areas. Snakes live just about anywhere; there are even sea snakes.

In most states, snakes are considered nongame wildlife and are protected by state laws, so don't indiscriminately kill them. You can construct a snake-proof fence around your house or a portion of your yard. It keeps out all venomous snakes and all but the most skilled nonvenomous climbers. The U.S. Fish and Wildlife Service gives the following instructions:

> The fence should be made of heavy galvanized hardware cloth, thirty-six inches wide with a quarter-inch mesh. The lower edge should be buried six inches in the ground, and the fence should be slanted outward from the bottom to the top at a thirty-degree angle. Place supporting stakes inside the fence and make sure that any gate is tightly fitted. Gates should be swung inward because of the outward slope of the fence. Any opening under the fence should be firmly filled—concrete is preferable. Tall vegetation just outside the fence should be kept cut, for snakes might use these plants to help climb over the fence.

Once you're familiar with different species in your area, you may develop the more tolerant attitude exhibited by those in the know.

Dave Sileck, beach activity director at Florida's Don Caesar Resort, says, "Black snakes are in the pool all the time. Usually they're just on their way to somewhere else. We just get the guests out of the pool—they enjoy anything different around the pool."

Generally, though, if you find snakes in your yard, just leave them alone. They are undoubtedly doing you a service by eating other critters that are truly undesirable—like mice and rats.

## Homebound Snakes

Snakes will come into a house. A damp, cool, dark basement full of accumulated treasures is attractive, so is an attic of mice and bats. You may feel more comfortable calling a professional: someone from the animal damage control staff, a wildlife official, or a private firm. If you're sure the snakes are not venomous, you can try removing them yourself with one of the following strategies.

Pile damp burlap bags in areas where you've seen the snakes, and cover each pile with a dry bag to hold in moisture. Snakes will think this is a great place to hide. Wait a couple of weeks, and then one afternoon when the snakes are resting among the bags, remove the bags with a large shovel. Put them into a large garbage bag and set the reptiles free outside. Be sure to prevent reentry by closing up any holes providing access to the house. Check the corners of doors and windows, and around the masonry, pipes, and electrical service entrances. If you use a mesh, be sure it's one-eighth-inch or smaller.

Residents of low-lying areas may meet their native terrestrial snakes during the wet season. These snakes are generally under a foot long and most are harmless. They can't climb trees, so they seek the dry shelter under the doors and in the cracks of houses. You may find them under the rug or in a dark closet, and you may find more than one. Since you can't just open the door and yell "Scat!" you'll have to capture and release the animals. Just gently sweep them into a waste-paper basket or other container and release them on high ground outside. Don't pick them up, because their bites could be painful. (Remember, even nonvenomous snakes do bite.) Then, weatherstrip the doors and fill in the holes.

James Knight, with the U.S. Department of Agriculture Extension

Service in New Mexico, hit upon a way to trap snakes. He attached six-by-twelve-inch sticky traps to sixteen-by-twenty-four-inch pieces of plywood with a hole drilled in one end and anchored the devices along walls in a dwelling. (Like most animals, snakes don't like to cross open spaces, so they'll tend to keep close to the walls.) When a snake would become stuck to the trap, usually after two days or so, he would retrieve the trap by slipping a hook (long handled!) through the hole drilled into the plywood. If you try this method, be sure not to place the traps near pipes or structures that can give leverage to a snake trying to escape.

After securing the captured snake, Knight pours cooking oil over the animal and gently pries it from the trap with a stick or pole. (Cooking oil will free animals affected by sticky traps and glues.) Knight had observed snakes in his lab for two months to make sure the method didn't harm the animals. Snakes shed any sticky residue with their old skins and suffered no ill effects.

No need to mention that if you're afraid of snakes, this is no method for you. It's an effective, nonlethal method provided you check the traps regularly and remove them entirely after a few weeks.

## Illegal Aliens

Most snakes that you encounter in the wild are small and not dangerous, but in southern Florida something else is happening. Owners of large, exotic snakes are letting their pets go free in the wild. Other pets, longing for the freedom of the wild, escape their owners. Boa constrictors and pythons are the favorite species for snake owners, and consequently, they are what you may encounter during a picnic. In June 1990 three exotic snakes were captured in Florida: a ten-, a twelve-, and a fifteen-footer. Nobody knows for sure, but hundreds of these snakes may be loose. And like every animal, they breed. Twelve-to-fifteen-foot pythons have no difficulty devouring a small dog. That's often the reason why snake owners release their pets: In the beginning, these snakes are cute (well, they are to their owners); later, feeding a fifteen-foot python can become expensive. You run out of poodles, or whatever, quickly.

Unfortunately, exotics like boas and pythons "are probably going to become a staple around here" according to Bill Zeigler. The exotic pet trade is big business with few restrictions; a lot of animals come into our country but aren't tracked or licensed. And many of the tropical reptiles thrive in south Florida's climate.

Zeigler says you'll probably find boas and pythons, both nonvenomous constrictors, around urban areas. They're there because we're there, and because we always attract plenty of rodents. While boas don't grow much over nine feet long, pythons can reach twenty feet. At that size, they're a danger to house pets and even small children. Pythons generally won't eat pets, but will squeeze them until they suffocate. So don't leave your kid napping alone in the yard because "you'll be stretching your luck," says Zeigler.

In general, it's a bad idea to let your pets roam. Not only does Fluffy make a litterbox of your neighbor's garden, she could become prey to any of a number of wandering predators. With the increasing frequency of python sightings in south Florida, there's even more trouble for unattended pets. But pythons or even poisonous snakes are less of a threat than other carnivores like coyotes.

## ❖ SUBVERTING SNAKES

1. Get rid of the junk or move it away from your high traffic areas. All sorts of animals will live in old appliances, cars, boxes, and other large cast-offs. If you can't cart them away, at least move them away from the areas where people walk, play, and congregate.

2. Raise the woodpile.

3. Get a cat to help you with the mouse problem.

4. Block access to your house, crawl spaces, and basements.

# BATS

Bats are good. Let me get my bias in the open right away. Bats eat mosquitoes, moths, small rodents, and other unpleasant creatures. A typical female bat will kill around six hundred mosquitoes and other insects in one evening. It's estimated that the bats around San Antonio, Texas, eat one million pounds of insects *every night*. Now that's some insecticide!

The problem with bats is that they have suffered a bum rap. We've harbored a lot of mistaken notions about bats for a long time. But here are the facts:

• Bats are rarely rabid. Estimates say that only one-half of one percent carry rabies. You have a greater chance of contracting rabies from a cow.

• Bats aren't blind. Not only do they actually see quite well, but in addition, many bats use a sophisticated sonar system that helps them locate their prey and zap it at high speed.

• Bats don't fly into people's hair. They're more scared of us than we are of them, and their sonar prevents them from making aerial mistakes like heading into a head of hair.

• The diseases bats carry (if any) are difficult for people to get. There's a possibility of picking op histoplasmosis, an airborne disease caused by microscopic funguses. The fungus often occurs in soil enriched with bat or bird excrement or in the excrement itself.

• Bats don't bite people (unless they're being handled), and they don't suck blood. Although there is a variety called vampire bats, they only suck the blood of livestock, and no vampire bat can be found in North America.

Bats live just about everywhere, except for the polar regions. Bats have been tracked at altitudes as high as ten thousand feet, about the elevation of the White Cloud Mountains in Idaho. They generally fly in the slow lane, although some bats can fly thirty or even sixty miles an hour.

Bats basically evolved along two paths—fruit- and nectar-eating and insect-eating, although only about fifteen percent are fruit eaters. Nectar- and fruit-eating bats are farmers of a sort. They drop seed-laden scat throughout the Americas, and nectar-eating bats help pollinate tropical flowers. Without bats it's possible that many species of plants would disappear. In fact, if it weren't for bats, you could forget about ever enjoying a margarita or tequila sunrise, for bats pollinate agave, a tropical plant used in the production of tequila. As a matter of fact, Bat Conservation International reports that bats pollinate or disperse seeds for over three hundred species in Old World tropics. You won't find any nectar- or fruit-eating bats in the United States, but we have forty-two varieties of our own bats. The most common is the little brown bat. Actually a cute little devil.

Insect eaters are the type that's prevalent here. These bats survive on a diet of bugs we love to hate: mosquitoes, caddis flies, moths, and beetles. To hone in on these tiny morsels—and to navigate at dusk, dawn, and occasionally at night, when insects are active—bats have evolved an adjunct to their sight. Their ace in the hole is the ability to echolocate, that is, use sound to see even in pitch darkness.

They bounce high-pitched squeaks (inaudible to most of us most of the time) off an unsuspecting insect, capture the return echoes with their oversized ears, and then perform awesome calculations to arrive precisely at the same spot in the same moment as the prey. Any bat sounds we may be able to hear are the audible squeakings, chatterings, and clicks they use to communicate among themselves. The bats capture the bugs in their wings and then transport them to their mouths.

The Greek name for the bat order *Chiroptera*, means "winged hands." Their arm and finger bones are greatly elongated to support the elastic skin and muscles they use to fly. Bats are the only mammals that really fly; flying squirrels are actually gliders.

Most bats we encounter are social and live in colonies that sometimes number in the thousands. The Mexican free-tailed bat, found in Mexico and caves in the American southwest, may form colonies of over a million.

Most bats mate in the late summer and fall but don't give birth until the spring. During the summer, the female bats and their offspring gather in large maternity colonies. (The males usually live solitary lives while the young are nursing, and some species form bachelor colonies. The stray bat you find behind the shutter is probably a male.)

Most mother bats have one offspring per year, who weighs about one-third of the mother's weight (that's like a 130-pound woman having a 43-pound baby). Mothers and babies keep the nursery temperature cranked on high—temperatures can reach over 130 degrees Fahrenheit—because the heat stimulates growth. After three weeks, the baby bat is ready to fly.

Mother bats are very devoted to their children. The only time mothers leave their babies is to go food shopping at night. What is amazing—at least to us—is that when they return to their cave, mother bats are able to find their own offspring among all the other bats. Once the young bats are weaned in late summer, the nursery disperses.

If these maternity colonies are disturbed by a human or other intruder, the effect can be devastating. Batmoms may be forced to abandon the offspring or may try to transport them, with deadly results. If the nursery is disturbed, the mothers will attempt to move the young by attaching them to their teats. Vandals—the human kind—will often kill the harmless young bats, and the disturbed hibernating bats forced into the cold air use up all their stored calories trying to keep warm, and die. Some of the forty-two species of bats in the United States are endangered, so be careful around potential bat habitats.

If the baby remains unmolested, it matures quickly and is soon flying and hunting with the rest.

Insect-eating bats emerge from their roosts in the evening and usually head straight for water. They drink by skimming the surface. Then the bats feed on insects for about half an hour, eating until their stomachs are distended with the fare. Then they rest and digest the meal. Mothers will return to the roost to feed their young, but others will find a temporary open roost under a porch, bridge, or tree. They'll feed again before dawn.

Although bats are mobile at night, they're vulnerable during the day—as well as during periods of hibernation—when they roost in caves, trees, or attics. They're prey to many animals: Owls, hawks, cats, raccoons, and snakes will eat them. As I alluded to earlier, vandalism and repeated disturbance of roosting caves is a primary cause for a drastic decline in bat population numbers. Forty percent of our native bats are on the endangered species list or are official candidates for it.

Despite the hazards, bats live a long time for wild animals, about thirty years—a result of their energy conservation adaptations. Active bats have a body temperature of around 100 degrees Fahrenheit. Roosting bats, unless they're in the nursery, save a little energy by dropping their body temperature to equal that of the environment, thus conserving energy. To survive cold winters, bats either hibernate or migrate to warmer climates.

## Attract Your Own Bats

While much of this book is devoted to telling you how to get rid of creatures, let me say a few words about attracting bats—should you want to rid your neighborhood of mosquitoes and moths. A family of bats is going to eat more mosquitoes in an evening than the most effective bug zapper can kill. (Besides, bug zappers, which rely on light to attract insects, actually bring insects into your yard, and they kill predator insects that prey on pests.)

The simplest, least expensive insecticide is a bat house. Bat houses, which look like oversized bluebird houses, can be purchased in most nature and wild-bird supply stores, as well as through a number of catalogs. Members of the free-tailed bat genus, found in the American southeast and southwest, prefer large structures placed high above the ground. Some of the successful ones are thirty feet above the ground and can accommodate a thousand bats. Bat Conservation International has good information on bats and bat houses; look in the appendix of this book for leads on BCI and other sources for bat houses. Unlike some birdhouses, bat houses are maintenance-free with an open bottom to allow droppings to fall away. The inside is partitioned to create crevices for thirty or so bats to roost in.

Since bats need water, bat houses near a permanent water source are the most likely to attract bats. Hang your bat house ten or twenty-five feet above the ground in a spot sheltered from the wind and facing south or southeast. If there are already occupied bat houses in the area, new ones have a better chance of succeeding. The bats need direct access to fly into the entry hole, so make sure it's not blocked by vegetation or wires. Bat houses attached to human houses are most successful because of the temperature stability.

Unfortunately, it may take up to a year and a half, or even several years, for bats to occupy the house. John Hadidian of the Center for Urban Wildlife of the National Park Service offers this hint: "You get bat guano from a place where bats have been known to den, and that will attract bats." He says people often put bat houses up when they want to get bats to move out of their attic, but they still want bats around. Once the bat house is erected, put some of the droppings from the attic in and around the bat house, Hadidian suggests. (Always wear gloves and a face mask while working with bat feces. Be sure to wash afterwards.) Bat pellets are about a quarter of an inch long and have a dark brown, gray, or black color. You might not know your bat house is occupied until you spot the pellets on the ground.

You may have more luck if you hang the house in the fall or winter so that bats will discover the roost when they become active in the spring. If after a couple of years you have no bat residents, try relocating the house. You can't introduce a captured or stunned bat to your bat house; their homing instincts to rejoin their kin are too strong to accept the substitute roost.

Most bat houses will eventually attract a small number of bats. However, multiply each bat times six hundred mosquitoes, and that's a powerful solution to the insect problem. Don't worry, by the way, that your yard might run out of flying insects, causing the bats to starve. First, it won't. And second, bats will travel up to fifteen miles each evening to eat.

Most of the time bats will occupy the bat house during warm weather and go elsewhere in the winter. Which is okay, because it's during the summertime that you want bats around.

While a bat house won't entice bats to leave your attic, don't worry that it will introduce them to it either. If your dwelling is attractive, they will already be there. If bats are in your house, maybe you should erect a suitable bat house before attempting to evict your unwanted tenants.

## Bats Out of the Belfry

Like other animals, bats are running out of suitable habitats—forests or loosely constructed outbuildings—which increasingly brings them into contact with humans. You like your house, why shouldn't bats? In fact, during World War II, U.S. military experts were so sure that bats would seek out roosts in homes, that they drafted the little critters. In project X-ray, Mexican free-tailed bats were equipped with small incendiary time bombs, refrigerated (to calm them down), and then dropped over enemy territory, particularly Japan. Like so many projects, the military abandoned this one too.

Bats are most likely to bother people in August, when infant bats are just starting to explore the world outside their cave (or tree, or attic). I can appreciate the point of view of people who would rather not have bats around, especially in their attic. Indoor bats are scary and leave scat all over the place. An accumulation of bat dung can be toxic, occasionally causing severe respiratory problems. People have devised a variety of methods for getting rid of nuisance bats. Recently, for example, about three dozen bats got trapped in a Salt Lake City, Utah, office building. A security guard herded the bats down twelve flights of stairs using a broom.

That's one way to get rid of them. Others include loud music, a stiff breeze, mothballs, and bright lights. Use five pounds of mothballs for about two thousand cubic feet; you can hang them in mesh bags or spread them on the floor and in the walls. Arthur Greenhall with the U.S. Fish and Wildlife Service reports that some people have successfully used a dog whistle hooked to oxygen cylinders or large aquarium pumps to repel bats. But ultrasonic devices have had little effect, he says. Bats can get used to about anything, so these solutions are iffy. Still, plenty of bats live in our homes without ever causing problems or even being detected.

In most states, bats are considered a nongame species so you can't hunt them, and poison is also out in most states. Besides, the problem is that you're providing a nice bat home, so new bats might eventually replace the old, dead bats. Some bats are protected by state or federal endangered species laws and, at any rate, officials frown upon the killing of any bat.

As a matter of fact, the University of Arizona in Tucson is facing charges related to killing bats. When they decided to get rid of a colony of bats roosting in the concrete football stadium, university personnel sprayed the roosting mammals with carbon dioxide from a fire extinguisher, both freezing and suffocating the animals. Then a local pest control operator followed up by applying a sticky bird repellent containing polybutene to the roosting surfaces, and the remaining bats had their wings glued to their bodies. The pest control company was fined by the state for improper use of a pesticide.

Last summer my wife, Peggy, along with my brother-in-law's family rented a house in Bolton Landing, New York, along the shore of Lake George. It was a pleasant house, devoid of air-conditioning, but who needs that kind of thing in the Adirondacks anyway. One of the alternatives to air-conditioning is leaving windows open, and one evening, somehow, a bat flew inside the house.

I have to admit that when I saw the bat swooping directly at me while I was playing with our ten-month-old child, I screeched a little: "There's a bat in the house!" Peggy, who thought I said, "There's a *man* in the house" came running out of the bedroom. When I explained what I had actually said, she scooped up the baby and ran back into the bedroom.

Here's what I did: Knowing that bats prefer the dark, I turned on all the lights in the house except for in one second-floor room. A couple of minutes later, the bat settled on the ceiling of that room. I immediately shut the door and ran downstairs. I then picked up the ham radio that we had been using to maintain communications while traveling around the lake and called for my brother-in-law Richard. "N3JAV calling N3JAY with an important message. Come in Rich." From the Green Mountains came a weak signal: "I'm here Bill."

"I just want to let you know so that you aren't surprised when you arrive, that there's a bat in the house. I've trapped him in one of the upstairs rooms." I told Rich that the bat was safely tucked away, but that I couldn't guarantee that it would stay there—not that the bat could open the door, but that it might squeeze out through the cracks.

A half hour later, Richard and his wife, Joanna, returned. Richard went upstairs. When he came back down, he reported, "Let me tell you something about the bat. There's a pair of tiny claws sticking out from beneath the door." Three of us were designated bat catchers—the other got to watch the babies. Wearing hats and windbreakers, and carrying flashlights, a broom, a dustpan, and a blanket, we gathered in front of the bat room. There we stood for a few minutes, gathering our strength, until one of us, Peggy, opened the door and dashed in. Reluctantly, Richard and I followed. We shut the door behind us, and immediately mimicked dolls at an amusement park's shooting gallery—ducking and bobbing our heads as the terrified bat swooped around the room, looking for a way to escape. We quickly arrived at a solution: Open the window and give the bat a way to fly out. Unfortunately, while the window was easy to open, removing the screen took about ten minutes. While Peggy worked at the screen, Richard and I covered her, by shining the flash light at the bat and swinging the broom. Finally, we got the screen off.

But the bat just kept flying around the room, apparently uninterested in the insect-laden outdoors. The three of us agreed—not a very smart bat. Meanwhile, we continued to duck and cover as the bat explored the space around the room—in particular the area around head-height. All this flying was getting the bat tired, and finally Peggy was able to toss a blanket over the bat. She then unfurled the blanket out the window and the bat took off like, well, like bats do.

The moral of this particular story is that if you find a bat in your house, don't worry. Bats are singularly uninterested in biting or even landing on you. Just make sure you don't forget the one thing we forgot: the video camera.

The best way to keep bats out of your home is to eliminate any openings through which they might enter. Unlike other mammals, such as squirrels and raccoons, bats won't chew their way into a house, but they will squeeze through the tiniest of holes, so you need to be diligent in finding and eliminating such entries.

Patience plays a role in getting rid of bats. As night begins to fall, stake out a comfortable spot to observe the bats as they exit your building, then block the entry holes. Becca Schad, owner of Wildlife Matters, a Virginia pest management firm says, "A lot of times in the summertime I work at night because that's when the bats are active. I go out to a house and start looking around about fifteen minutes or so before it starts getting dark and the bats become active and fly out. By noting where their exits are, I can go behind them and close up the exits."

Sounds easy, right? Blocking up the easiest bat access doesn't mean they won't find another hole. So before you block the active access, close off the potential ones. One way to find potential bat doors is to seek out places where the air flows from the house. You can find air leaks with inexpensive household items: incense, tissue paper and a hanger, or a candle. Tape the tissue paper along the bottom stem of a clothes hanger and hold the device near eaves, walls, and windows you suspect are leaking air. Smoke flow from incense or a flickering flame will indicate the same.

Some spots are obvious. Schad says that bats often enter through the louvered vents in an attic and recommends using one-quarter-inch hardware cloth to block the access. Vents, spaces around doors and windows, loose screens, chimneys, and cracked flashing are other likely spaces.

The best time to bat-proof is at the end of the summer when young bats are ready to fly and they're not yet ready to hibernate. After you close up the auxiliary holes, block the main exit. After you block up the main exit, open it on several successive nights to make sure all the bats got out. Repeat as necessary.

If it's awkward to close off the final exit after dark, there are a number of one-way doors that can be used temporarily until the bats are out. Some one-way doors collapse after the bats fly out and others use a cone design where bats exit from the small end of the cone and can't fly back in. Perhaps one of the easiest temporary one-way doors is bird netting strung across the main access. The bats leaving the building will climb down the netting, drop, and fly. On returning, they would usually fly right into the hole, but now they can't because of the netting. You can make permanent repairs with caulk, weatherstripping, metal flashing, screens, or insulation. You can block holes under roofing with fiberglass batting or rustproof scouring pads.

Schad adds, "More and more these days people are realizing that bats really don't deserve their horrible reputation and that they really are beneficial because they eat insects." Even so, a lot of people are still pretty squeamish about actually going up in an attic where they know there are bats, so "a lot of times they call me." Yet, she says, bats "really are one of the more harmless creatures that get into houses."

Sometimes people discover bats hibernating in their attic over the winter. Don't disturb the bats then, because you can't remove hibernating bats without endangering them. If you oust a bat in the winter, it may well die of starvation because its insect prey is no longer present. Also, bats awakened from their winter hibernation will burn off their stored fat reserve trying to keep warm, and may not be able to make it until spring.

Another reason not to remove bats in winter is that even if you did wake a few up and get them out, you couldn't be sure that you had gotten them all; bats are good at hiding in the cracks and crevices.

Also, avoid sealing off your attic when bats are hibernating inside, because when the bats become active in the springtime, they'll be trapped inside and die (or they may try to make their way down into the rest of your house).

Hibernating bats won't hurt anything. Guano won't accumulate, and the pest insects that come along with bats (bat fleas, ticks, mites, and bedbugs) rarely bite humans. So hold out until spring, and don't disturb the animals when you go up to the attic after your skis. Then, one fine, spring evening watch the innocent little animals as they flutter into the sunset. And then scurry upstairs to prevent them from returning.

If you need to clean up after some bat visitors, first dampen guano with water before removing it. You should also wear a face mask or dust respirator while you're doing the work. Spores in the guano (or in chicken or bird droppings) may cause histoplasmosis, a disease that, while rarely serious, usually causes headache, fever, and sometimes a cough and chest pain. The disease is cured with antifungal medication.

If you fail at excluding, harassing, and evicting bats, you may have to trap them, but that's a case for an expert. People bitten by bats are treated for rabies, so it's really best to have a professional trapper take care of the job.

Bats in your living space are a different problem. They're panicked. You're panicked. Bad things could happen. You need to be level-headed. If it's night, turn out all the lights, open the windows and doors and let the bat fly out. Try to isolate it in one room by closing off the rest of the house. Failing that, when the bat lands, try to cover it with a coffee can or cardboard tube; it will probably be grateful for the cover. Slip cardboard over the container opening and release the bat outside. You can try to catch a flying bat with a butterfly net, in a towel, or with leather-gloved hands, but remember that even the gentle bat may bite in self-defense.

## ❖ BAFFLING BATS

1. Cover louvered vents in attic with quarter-inch hardware cloth.

2. Block up any other holes in your home, especially near the roof and eaves.

3. Give bats alternatives to your home—bat houses—when you oust them.

4. Don't disturb hibernating bats in winter.

5. Make sure you don't trap bats in the house when you attempt to exclude them; they're great at hide-and-seek.

# BIG ANIMALS: BEARS, MOOSE, AND SEA LIONS

## BEARS

Smokey the bear. Yogi, who's smarter than your average bear. Your childhood teddy bear. Gentle Ben. In children's literature, and in adults' hearts, the bear evokes warm, contented emotions. There are 800-number services that deliver teddy bears anywhere in the country. For nearly everyone, bears evoke pleasant childhood memories.

Bears are smart, as Yogi would insist. Just as smart, for example, as the hunter who was prowling around Iron River, Wisconsin, in the fall of 1990. Stalking deer, this hunter was suddenly attacked by a black bear, who picked the hunter up and "swatted me like a badminton birdie." The hunter survived. When the local game warden was asked why the animal did this, the warden said that the bear was probably protecting its food or was startled. I offer another explanation: The bear thought, "Today deer hunting; tomorrow, bears are going to be the target." Well, we'll never know for certain.

When bears become nuisances to people, often they are identified and moved. This is what happened to two black bears who were roaming around the Lower Paxton Turnpike in Pennsylvania—and it is what happens to hundreds of bears a year. Usually, in fact almost always, people are the reason that bears become a problem. Like most animals, bears prefer to avoid humans (how smart they are!), but when given handouts, or allowed to rummage through our garbage, they often become accustomed to human habitation. Then they become dangerous.

My sister was lucky with a bear encounter. Diane was relaxing by a river in California when a bear swooped by and grabbed her knapsack, complete with lunch. She yelled "Bear!" and a fisherman downriver hearing her warning moved faster than any other human she's seen. The fisherman was unharmed, but Diane never saw her pack again.

There are three North American bears: the black, brown, and polar bears. From here it gets confusing. Black bears (rarely are they black) have the widest distribution, and they're found throughout Canada and along the United States–Canada border, in wide swatches down the East and West Coasts of the United States and in the Rockies. Brown bears and their subspecies, the grizzly and Kodiak brown bear, are found in the Northwest. Some biologists don't differentiate among the three and refer to them all as grizzlies. Polar bears are only found in the far northwestern reaches of the continent, and I won't cover them in this chapter.

As the name implies, brown bears are brown, ranging in color from yellowish to dark brown. They may be three and a half feet tall at the shoulder and weigh up to eight hundred pounds. Grizzlies have a hump on their backs, a muscle mass to help them dig out marmots and other ground-dwelling animals. Many grizzlies have a grizzled appearance—their hairs are tipped with gray or they have light and dark hair interspersed in their coats. It takes time to learn to identify the grizzly, a big brown bear won't do—it might be a big black bear, an animal that shares the same habitat but behaves differently than the grizzly. The grizzly's average weight is three to four hundred pounds, although the male, unlike the female, continues to grow throughout its life. One ten-year-old male weighed over nine hundred pounds. The grizzly is identified by the shoulder hump, a concave face, and long front claws.

Despite their name, black bears range in color from reddish to cinnamon or gray. Black bears weigh from two to three hundred pounds with six hundred pounds representing the extreme. Black bears have shorter claws than a grizzly, lack a shoulder hump, and have a straight or "Roman" profile. Often black bears appear to have humps, usually when they have their heads down below their shoulders as they feed.

Bears live solitary lives except for when a mother is raising young or when a male and female pair up for a month or so in the summer to mate. Even bears at the same food source generally avoid contact with one another. A bear lives in a loosely defined home range that may overlap with that of another bear, and a bear will travel throughout his range harvesting food in season. Males have larger home ranges than females, varying in size from five to fifteen miles in diameter. They use different parts of their ranges depending on the season, and while they don't keep one den to return to every night, for bears that live in climes requiring it, the home range contains the den for winter hibernation.

Bears hibernate in crevices in rocks or under stumps or logs, in caves, in holes in the ground or even under buildings. Dens offer shelter from the elements, and they can't be wet. On occasion, bears will hibernate in heavy brush. Naturally, bears in the southern regions don't hibernate.

During the late summer and fall, bears put on weight for hibernation, building up a thick layer of fat, and enter their dens between October and January. Before denning, bears become less active and don't eat much. A plug of material forms in the lower colon; the bear does not defecate during hibernation. During hibernation, the bear's body temperature drops about forty degrees Fahrenheit, the heart rate slows, and the metabolism drops by about half. A bear may lose up to twenty percent of its body weight during hibernation. Between March and May, the bear emerges from hibernation, dislodges the fecal plug, and slowly resumes bearlike activities.

Bears are most active at dawn and dusk. They usually sleep at night, in dense brush, but when food is plentiful, they may be active all day and night. Black bears are mostly vegetarian, but grizzlies eat more meat than fruit and nuts. Grizzlies are expert fishers and will also dig into the ground after tubers, bulbs, and roots, and ground squirrels. Black bears eat grasses and emerging plants in the spring, tree and shrub fruits and berries in summer, and berries, larger fruits, and nuts like acorns and beechnuts in the fall. Both also eat insects, especially ants, wasps, honeybees (and honey), grasshoppers, grubs, mice, voles, and carrion. Bears may be destructive feeders, ripping apart stumps or logs to get at insects or they may strip bark off evergreens to get at inner bark. All bears will eat garbage, because the food found there is easily turned into energy their simple digestive systems can use.

In late spring to early summer, a male bear will begin to follow a female and the two will associate for almost a month as they court and mate. Other males may challenge the mated male's right to breed, and fighting may occur. After mating occurs, in late June or early July, the couple parts. Bears have delayed implantation, so although the egg is fertilized, it is not implanted in the uterus until around November. Gestation is short—six to eight weeks, and cubs born in the winter den weigh only six to ten ounces. Although they're born in January or February when the mother is semidormant, they do receive the minimal care they need. They are blind and nearly furless and they nurse and sleep as the mother sleeps. Although first-time mothers have one cub, twins are the norm thereafter.

A few weeks after birth, cubs have fur, and after forty days, they open their eyes. Between March and May, depending on the temperature, the cubs, weighing about five pounds, emerge from the den with their mother. Cubs stay with mother through summer and into fall, by which time they weigh about fifty-five pounds. In early winter, the mother finds a den and she and her cubs den together. The next spring, the mother is ready to mate again, and either she or her mate drive the young off. Cubs may stay together for another year, but bears become sexually mature at three and a half years, and will reproduce a few years later. Bears live twelve to fifteen years.

## Grizzly Bear

The grizzly bear suffers from "a public relations problem," says one U.S. Fish and Wildlife Service officer. In 1975 this large, ferocious carnivore—*Ursus arctos horribilis*—joined the endangered species list. Before we settled North America, the grizzly, whose range extended from northern Mexico to the Arctic Circle, numbered about 100,000. Today, there are an estimated one thousand bears left. Declining habitat has been a principle reason why the grizzly is nearly gone, but predation has taken its toll: Hunted for fur, fear, and sport, the grizzly has never been a match for a rifle.

In the United States, grizzlies make their home in Alaska, the northern Cascades region of Washington state, the Cabinet-Yaak part of Montana, the northern Continental Divide area of Montana, and, of course, Yellowstone National Park. Grizzlies are rumored to be roaming the San Juan Mountains of Colorado and the Bitterroot Mountains of Idaho.

The U.S. Fish and Wildlife Service has begun a program of reintroducing the grizzly into certain parts of the country, including the San Juan Mountains. Hikers and farmers probably shouldn't expect any problems from these bears because there are only so many bears available to be relocated. Grizzlies reproduce slowly in the wild and not at all in captivity, making them a difficult candidate for repopulation efforts. The government's transplant plans meet resistance whenever they are announced: Many people fear bears. Indeed, grizzlies are dangerous. Only where people are few can grizzlies be reintroduced. Grizzlies thrive best in remote areas that provide the seclusion they need for raising families.

## Black Bears

While the grizzly is the one with the public image problem, the black bear is no teddy bear. They have adapted a little better to living near humans and therefore have held onto their traditional range better. Where their range overlaps with that of the grizzly, the black bear usually contents itself with the lowland during the summer months (which is often where people live), because the grizzly will drive the black bear out of the high ranges.

In his book *Bear Attacks: Their Causes and Avoidances,* Stephen Herrero found twenty deaths attributable to black bears between 1900 and 1980, and predation appeared to be the cause in ninety percent of the cases. By contrast, the grizzly attacks he studied over the same time period resulted in nineteen deaths, mostly when human hikers surprised a grizzly or a grizzly and her cubs. To put it in perspective, you have a greater chance of being hit by lightning than being killed by a bear.

Black bears are adapted to live primarily in forested areas, while grizzlies live in more open areas. The black bear, less apt to fight aggressively than the grizzly, depends on the cover for survival, hiding in the brush and climbing trees for protection. Still, as human populations threaten their traditional habitat, more grizzlies are moving into the forest and outlying areas.

## Bear Encounters

All bears are seeing more people than they'd like. Encounters on your turf are different than ones in the backcountry. If you can, retreat to your home or a hard-shelled vehicle if you encounter a bear outside, but keep in mind that they run faster than any Olympic sprinter, so you need a good lead. If you're sure the bear isn't a grizzly and it hasn't already found a food source, you may be able to bluff it away with noise and by waving your hands in the air. You'll never bluff a grizzly, so don't even consider it.

Bears aren't necessarily nosing around your house looking for a handout. "Bears are curious, they'll show up on your porch and look in the windows," says Heidi Youmans with the Montana State Fish, Wildlife, and Parks Department. But don't give them reason to return: Keep a sanitary yard. If your garbage is smelly or you have any food sources outside, bears will come around for an easy meal—and once they're in the yard, they'll explore other food sources around the house. Keep garbage inside until it's time to take it to the dump. Protect your gardens and orchards with an electric fence—and get a big, well-trained dog.

Even living near a landfill or having sloppy neighbors is a cause for concern if it brings bears down around your dwelling. Once they learn about garbage, they seek it out and have an uncanny ability to find it anywhere. To discourage bears from coming around your house, keep brush cut down so bears don't have cover to get close. Four walls and a roof may not be enough to stop bears that have associated the home with food; bears have torn through walls, roofs, doors, and windows in search of food.

Don't leave your children playing in the yard. Fifty percent of the deaths from black bears in Herrero's study were people under age eighteen. In one case, a black bear attacked and partially ate a child who was playing near her home. If a bear is hanging around your house, you may be able to startle it away with a gas exploder, a shotgun blast, loud music, or flashing lights. In time, however, the animal will get used to the noise and lights, and the methods will become ineffective—then the animal will be more dangerous than ever. States regulate the taking of bears, so check with officials before you kill a problem bear.

In the late 1980s scientists started looking for an effective bear repellent. They hit upon hot pepper spray, a mace for bears. It was tested once in 1984 by a field technician studying grizzlies. The technician got too close to a bear he was tracking so the bear charged him. The man sprayed the bear, which backed off, but then charged again and bit the man, who used the spray again. The bear retreated. It might be a good idea to keep a can of the mace at your side if you have to be away from protective shelter in bear country. You can get the spray at many outdoor stores.

If you live in bear country, get information from the local wildlife agencies about the animals and what to do when you encounter them. Most of us can't interpret the bear's signals, and that makes it hard to know what to do when we see one. Local wildlife officials are familiar with any problem bears and can best advise you. A book I mentioned earlier will give you a good picture of bear biology and how to react in a bear encounter: *Bear Attacks: Their Causes and Avoidance,* by Stephen Herrero. Although he focuses mostly on backcountry encounters, Herrero, a bear biologist who teaches at the University of Calgary, looks at problems around the home, too.

If you're backpacking and take along food, also take common sense. If you don't want to risk your tent suddenly having an open door, then hang your food high and far away. All your food. And soap, toothpaste, toothbrush, deodorant, and eating utensils. Just because the items you're carrying don't appear to have an aroma to you, doesn't mean that bears can't smell them Anything fragrant or oily is attractive.

How high and how far? The farther away the better. You don't want a frustrated bear to decide to inspect some other nearby man-made object. Besides, the farther away your food cache is, the more time you'll have to investigate that strange noise-in-the-night. As for how high—well, to give a correct answer, you have to take into account the fact that *other* animals are interested in your food, too. Hang you food at least six feet up and out on the thinnest branch that will support the weight. Remember, animals in the woods have nothing better to do with their time than to try to get to your food.

## ❖ BEING SMARTER THAN THE AVERAGE BEAR

1. If you live in bear country, learn about bear habits. Get in touch with wildlife officials to find out about any problem areas near you.

2. Don't leave small children playing in openings close to forest and brush cover where bears can hide.

3. Keep garbage secured in an outbuilding or your house until you can take it away. Make sure any garbage you burn is completely consumed.

## MOOSE

Moose are from the same family as the white-tailed and mule deer, but they are considerably larger. The males weigh between nine and fourteen hundred pounds and can be up to seven feet tall at the shoulder. Females are somewhat smaller at seven to eleven hundred pounds. Males have large, flat antlers. Their coat color ranges from tan to nearly black, and both sexes have a dewlap, a long flap of skin hanging under the chin. Moose are found in Canada, Alaska, the Rockies, around the Great Lakes, and in New England.

They're active at most times of the day, alternating between feeding and sleeping. Moose are vegetarians, and like deer, have four stomachs and chew their cuds. They're fond of aquatic vegetation, and you'll often find moose wading in water to eat it; in fact they may completely submerge for aquatic vegetation. They also eat ground plants, and will push saplings down with their bodies to eat the tender tops. They prefer willows, gray and white birch, quaking aspen, balsam poplar, and balsam fir. In the winter, they eat the twigs and bark of deciduous trees, conifers, and shrubs, and, of course, lose weight because their diet is so limited. Some estimates say moose lose up to half their body weight over the winter. They do have an advantage for winter survival; they have extraordinarily long legs and can walk through deep drifts.

Although they are large, they can run up to thirty-five miles an hour and are good swimmers too. They may swim a "short" distance of eight miles—perhaps from the mainland to an island.

Moose move around from season to season, so it's difficult to estimate the size of their home ranges. For instance they may move from mountaintops in summer to lowlands in winter. During any one season, their home ranges are probably two to four square miles.

In late fall—November to December—moose form in loose groups of up to twenty, groups of bulls, cows and their calves, and young moose. They stay together until spring, when the group breaks apart. Through the summer, bulls are solitary or with younger bulls, and they remain this way until fall when they are searching for mates.

In the autumn, the adult bulls are on the prowl looking for females, who are still with their previous year's calves. Females call out when they are in estrous, and are usually in open areas. When the bull hears a cow, he moves toward her, grunting. Once they finally find one another, they stay together one or two weeks, and then the male seeks another female and mates with her as well.

Males are very protective of their mating rights and don't allow other males to approach. If another male challenges a mated bull, there may be a fight, but first they engage in a ritual display. The bull will circle the intruder, hit shrubs with his antlers, sway from side to side, urinate in the dirt and roll in it. If this behavior doesn't drive off the challenger, the two may fight. Urinating in the dirt to create a muddy wallow is a common occurrence. The bull repeatedly urinates on the ground and scrapes the area to work the liquid and earth into a wet mess. Eventually he may lie in it and roll, and the female and calf may follow suit. Wallows may stimulate mating behavior.

The pregnant cow remains with her yearling through the winter but finally drives it away around June when, after an eight-month gestation, she is ready to give birth. She finds a protected spot where she produces her twenty-two-to-thirty-five-pound calf. Moose usually have one calf unless it has been an especially good year with plentiful food, and then twins are common. In good years, when they get plenty of food and stay healthy, yearling females can mate, but they have a single birth.

Calves remain hidden for two or three days. They're covered with reddish-brown, short woolly fur, and their long ears and legs contrast with their short bodies. Cows are very protective of their calves and will not let moose or other animals get near them. Calves weigh seventy-five pounds by nine weeks and aren't fully weaned until the cow's next calf is born the following summer. Moose hit their prime at eight to ten years, but may live twenty-three years.

Only the bulls have antlers, which are shed each year between December and March. The growth starts in the spring and continues through summer. The diameter of the antler base gets larger each year. Bull calves have only spikes first year, but yearlings and two-year-olds may have spikes or branched antlers.

Keep in mind that moose are big animals. That fact yields a single important recommendation: Keep out of their way.

There's not a whole lot that can be said about encounters with moose, because in almost all circumstances you want to keep as far away from a moose as you can. The moose will try to do the same thing, which is fortunate, because moose are very big, very strong, and very determined. They also have those large antlers, which they are not afraid to use. You can wander the woods for years and not see a single moose, but when you do spot one, hold your position and let the moose make the first move. If he stands still, take the opportunity to watch or shoot pictures. If he moves toward you, interpret that message correctly and back away. While moose can run pretty fast, they sometimes have trouble maneuvering. So if you're being pursued by an angry moose, zigzag wildly around trees—better yet, climb one.

## ❖ MEETING MOOSE

1. When you encounter a moose, give way, get off his path. If the moose approaches, retreat. Be careful in the fall mating season and spring birth season when moose are really edgy.

2. You're most likely to see hungry, desperate moose around your home in the winter when they'll decimate your trees and shrubs. Try coating target trees with deer repellents like Deer Away or Ropel.

3. Try feeding a lone moose through the winter; it will move on when the weather turns warm. Try root vegetables, tree boughs, or hay.

## SEA LIONS

In most California zoos sea lions are popular attractions. Their acrobatic antics and doglike barks amuse and amaze children and adults.

In some cities, while the zoo proudly displays these creatures, on the other side of town city officials desperately try to think of ways to rid their docks of sea lions. Why would anybody want to rid themselves of sea lions, which many people are willing to pay money to see? Well, in San Francisco and Monterey, for example, a proliferation of sea lions has, at times, created a nuisance. Looking for a place to rest, sea lions have decided to relax on boats; when a half-dozen eight hundred-pound sea lions climb on a pleasure boat, it sinks. Same thing for the docks themselves. Then there's all that sea lion fecal material, which is not insignificant, given the sea lions' size.

You can't kill sea lions because they are protected by the 1972 Marine Mammal Protection Act.

You can't move them either. When a tamed sea lion became trapped off the Shilshole Marina in Seattle, Washington, nothing could coax him into his cage. Shove as they might, neither the scientists nor animal handlers could push Sandy, this 915-pound animal, into his cage. He wouldn't move. Hours later Sandy simply decided that the cage wasn't so bad after all.

California sea lions, like the other sea lions, are intelligent and playful mammals. The California sea lion is most often seen in zoos and circuses, and can learn an array of tricks, although it's not any more intelligent than its kin.

Sea lions are dark brown and have thin, short, coarse hair. Their front limbs are flippers, and their hind limbs are fused into one large flipper that they can turn beneath them to locomote on land. Male California sea lions grow quite large: six feet long and up to six hundred pounds. Females are about the same length but weigh much less—about two hundred pounds. Sea lions feed exclusively on fish and squid.

They breed once a year, gathering on shore in harems of up to fifteen females dominated by a male.

If you can't move them, shoot them, or scare them away, what can you do? The only solution is to wait for sea lions to want to go away on their own.

## ❖ SURVIVING SEA LIONS

1. Moor your boat away from sea lion congregation areas.

2. Alter sunny, flat areas where sea lions gather; make the animals uncomfortable. Create shade, slippery slopes, and pointy, sharp surfaces.

3. Noise makers like gas exploders may startle the animals and encourage them to move on.

# BEAVERS

People know beavers as those industrious tree-felling, dam-building, lodge-living animals. They are the animal kingdom's model of the work ethic incarnate.

However, much as we might admire this behavior in principle, all of this industry can get to be a pain if it's being applied to your property. Aquatic engineers with minds of their own (including their own ideas about who holds title to the pond), beavers pay little attention to property lines and have no respect for your landscaping improvements.

It's hard to fault them, though. This behavior has been programmed into beavers for eons: Nature designed them to do exactly what they're doing. And designed them beautifully.

Except for dry areas in the West and most of Florida, beavers range throughout the United States. Weighing between thirty-five and fifty pounds, and reaching lengths of twenty-five to thirty inches (excluding their flat, scaly tails), beavers are our largest rodents. But unlike other rodents, beavers are especially adapted for a life in watery environments—living in or beside wooded lakes, ponds, streams, and other wetlands.

Beavers have some sort of neurotic need to stop running water, so stream-side homes are prime real estate for them. First they'll cut down trees and build dams to stop the water. Then they engineer the habitat more to their liking. They girdle the trees, killing them, and leave any remaining trees to be killed by the flooding caused by the dam. In time, water-loving and fast-growing vegetation begins to thrive. Willow, sweetgum, poplars, and buttonbush grow around the pond, creating a food source for the beavers.

Because they spend much of their time in the water, beavers have warm, thick reddish-brown fur (so luxurious that beaver pelts have been highly prized through the centuries by fur traders and fashion-makers). In addition, glands near the beaver's tail secret an oily sub-stance that helps "waterproof" the animal by keeping water away from the animal's skin. Other adaptations for the watery life include special nostrils and ears that automatically close when the animal submerges (they may stay underwater for as long as ten minutes), webbed hind feet, and flat, paddlelike tails. To communicate danger, they slap their tails against the water. (They also use their tails as a prop while sitting upright on land.)

Another beaver hallmark is their unusually long, sharp incisors, which are bright orange on the front. They use these to girdle and cut down trees. Because the teeth wear down under this heavy labor, they grow continuously throughout the beaver's life. These incisors are beveled on the back side to be continuously sharpened as the beaver gnaws and chews.

Trees and branches are all-purpose materials to beavers. Woven together across a stream, branches are used to form dams. Dams vary in size according to the topography of the land. A flatland beaver may build a short and long dam—about three feet high and a quarter-mile long. In hilly country the dams are higher and not as long. Beavers dam up waterways to slow the speed of the water and raise its level to an appropriate height for their lodges.

This brings us to the second use of trees: as building material for lodges. Beavers build lodges of large branches and logs and cover them with smaller vegetation and mud to hold it all together. The lodges may be built along the shore or in the middle of a body of water. (However, if a beaver chooses to live beside a swiftly flowing waterway, it may burrow into the bank instead of building a lodge.) Beavers are always improving their lodges, making them bigger and stronger. Typically, a lodge will be about four feet high and twenty feet wide.

After the beavers finish building the mound of vegetation, they tunnel in the underwater entrances and hollow out a cavity—above the water level—for the family to live in. There are at least two and up to four underwater entrances. The underwater entrances effectively keep out potential predators, like coyote and bobcat, although river otter and mink will swim up and attack kits.

Finally, along with using trees for damming and building materials, beavers use them as food. They often use their sharp incisors to fell slender trees and then feed on tender topmost branches. One beaver can fell a three-inch diameter tree in about ten minutes, a five-inch diameter tree in a half hour.

Rather ungainly on land, they don't venture out on foot more than a hundred yards or so away from their lodge to collect their food. But they may go as far as a half-mile in water, logging the far shores of a pond or a river, and dragging the branches back to their lodges through the water.

Beavers are especially fond of aspen, maple, alder, willow, birch, and sweet gum. In the West, they'll eat conifers like the Douglas fir and pine. They don't eat the inner wood of a tree, only bark, leaves, and twigs. For bark, they prefer that of aspen, cottonwood, balsam, poplar, apple, ash, maple, birch, alder, and willows. They'll girdle sweetgum and pines so they can eat the gum or storax that seeps out of the trees. They also enjoy some cultivated crops and will venture away from the safety of the lodge to gather corn or soybeans. Waste not want not; beavers incorporate uneaten parts of the booty into dams. They also feed on most herbaceous and some aquatic plants.

Beavers are generally nocturnal, active for about twelve hours a night. It's not unusual to see them during the day, though, since they begin activity in the late afternoon and go until early morning.

Beavers don't hibernate in winter and rely on food they store on the bottom of the pond; called a *cache*. They swim underwater, gather some branches from the cache and take them up into the lodge for feeding. They may also venture out in the winter to eat bark from nearby trees.

The dens and dams that beavers build are so labor-intensive that beavers live in colonies, which work cooperatively to build and maintain their structures and gather food. A colony usually consists of an adult male and female pair, the current season's offspring and the previous year's offspring. Female adult beavers are dominant; when they speak, all colony members listen. (This might account for the fact that male beavers are expected to help care for the kits, and do so, no questions asked.) Adult females sound the tail alarm to warn of approaching enemies.

Beavers are monogamous, mating throughout the year with one mate only. Beavers breed between January and March in the North and between November and January in the South. Gestation lasts about a hundred days, and both the male and the yearlings may be present when kits are born.

The kits, two or three, are born with fur and able to walk. In fact, they may go into the water after only a few days. By the time they're ten days old, they can dive into the water and swim about, and by two months, they leave the lodge with the parents. The parents introduce leaves to the kits' diet early on, and after a few months, the youngsters begin to feed on bark. By summer, they are pretty heavy, weighing ten to fifteen pounds. Their yearling siblings weigh fifteen to thirty pounds now.

By the time autumn rolls around, the entire family gets to work storing food for the winter and building or repairing the lodge and dam. During their second summer, the yearlings may wander away for a while, but they generally return for the winter to spend it with the family in the lodge. By their third summer, the young beavers are sexually mature and ready to start a colony of their own. They set off to find a home range and mate, probably the longest migration of their lives—possibly thirty miles. Beavers live about twenty-one years.

You might notice mounds of mud at the water's edge of a beaver habitat; that's their way of marking their territory. The mounds are scent-marked with castor, washed out of the beaver's castor glands by urine. There may be forty or a hundred mounds in a territory, with a heavier frequency around the dam and lodge.

## Don't Leave It to Beaver

Not everybody has a beaver problem, but some people's property is big enough to encompass ponds and streams where beavers live. In my opinion, we should all be so fortunate as to have a problem with beavers.

It's more like beavers have a people problem. Some animals clash with humans because they actively irritate—they tip garbage cans, gnaw through walls, or nest in attics. But the beaver is just trying to make a home, one probably not too close to human habitation.

The Fund for Animals points out several positive effects of beaver dams:

- The dams help with flood control by holding back water and releasing it at a slow rate.
- The wetlands created by the dams provide crucial habitat for other wildlife.
- The dams are effective for drought control, regulating a slow flow of water downstream.
- Beaver dams store water that can be used for irrigation or fighting fires.
- Backed up water recharges the groundwater aquifer.
- Dams filter water passing through them.
- Beaver are fun and easy to watch because they're not shy animals.

But a beaver doesn't live quietly in a tree; it alters the landscape when it makes a home. They've caused major damage in some areas; estimates range from three to five million dollars' worth of damage in the southeast from flooding and crop and timber loss. (In defense of beavers, though, you have to wonder where the caretakers of the damaged land were. A beaver pond doesn't spring up overnight!) Reservoir dams have been destroyed and trains have derailed because of damage done by burrowing beaver.

The average beaver-bothered homeowner has it easy by comparison—damaged trees and plants and minor flooding. But some homeowners aren't keen on beavers destroying their newly planted trees, especially aspens and willow. They also might not like the idea of backyard streams being dammed and flooded to form ponds and marshes.

If you want to keep beavers away from your trees, one thing you can do is fence your trees off with thick burlap. Wrap the material around the tree so the rotund rodent can't sink its teeth into the bark. If you use burlap you might fortify it with my favorite chemical weapon: cayenne pepper. Spray the burlap you've wrapped around the tree with water mixed with cayenne, or spray the tree directly. Some people have also had luck with the deer repellents like Ropel and Deer Away sprayed onto bark and foliage.

In a similar vein, Beaver Defenders, a New Jersey group, recommends heavy wire fences be placed six to twelve inches from the tree so the beavers can't touch their teeth to the tree. The wire has to be heavy enough to stand up to thirty or forty pounds of hungry beaver pressing against it. Also make sure it's firmly attached to the ground so the beaver can't crawl under it. You'll need to adjust the wires every few years to make sure you don't girdle the tree yourself. Sherri Tippie, who heads up the Colorado group Wildlife 2000, recommends concrete-reinforcing wire because she thinks it's not as visible.

Your fence should be three and a half or four feet high, because the beavers will stand up to do their gnawing, says Larry Manger of the U.S. Department of Agriculture in California.

Instead of fencing trees individually, you could install a metal fence to protect a larger area. (The fence won't keep raccoons or squirrels off the tree, but they aren't likely to cut down your tree, either.) Cutting is most likely to occur in mid to late fall or in the early spring. But if you try to fence in an area too close to a stream the beaver wants to dam, it might incorporate your fence into the structure.

While bold about cutting down trees, beavers are skittish about unnatural objects. So, near treestands that you want to protect you might suspend a thirty-six-inch square white flag between two poles to scare beavers away.

Expect damage to trees around the beaver's dam, especially to three that are their favorite foods. Beavers prefer smallish trees like the willow, cottonwood, and the poplar species, trees that regenerate new growth quickly. (Beavers actually stimulate growth in willows.) Often the trees we like aren't the first choice for chow to the beaver, but they're what's available. If a beaver does manage to fell a large tree, don't remove it immediately. Let the beavers strip it of edible material, and then turn the remains into firewood. If you take the tree away, the beavers will just have to cut down another one. Despite what years of cartoons have taught you, beavers don't plan which way a felled tree will fall. A classic example of being in the wrong place at the wrong time, a car was smashed when a beaver cut down a tree as the car was driving by. But that's the least of your worries. It could fall on a utility wire.

Leftovers from a felled tree will be used to block flowing water and flood an area. Beaver engineering is an inexact science. They don't figure out how much water can flow through their dams and still leave enough to protect their lodges; they stop all the water and let it back up where it may. They can't abide a leak in the dam and fix leaks immediately. That's when the problems start.

Wayne Pacelle, the national director of the Fund for Animals, understands that not everyone is willing to put up with a flooded yard or house so that a beaver family can have a nice pond. In fact, the organization offers technical assistance to folks who want to mitigate the flooding effects; clients include the U.S. Army and several state governments. A Canadian business, D.C.P. Consulting, has designed devices to let beaver dams drain but not drive off the beaver. Both organizations are listed in the resources section (pages 231–257).

Generally, the idea is to allow a beaver pond to drain the water in a way that keeps humans and beaver happy, reducing the size of the pond but not eliminating it. Beavers usually notice holes in their dams because they hear, see, or feel the running water. You have to drain the water in such a way that the beaver won't sense the movement—the method must be quiet, invisible, and gentle. Simple, right?

Maybe. The Fund recommends driving a pipe (PVC or other material) through the dam so the water can move downstream. First you have to figure out the mechanics so that it will drain the perfect amount of water; if you leave the beaver high and dry, it may build a dam elsewhere, somewhere even more inconvenient. Figure out the volume of water you need to remove, and install a pipe of the appropriate size. Now, the pipe needs to be quiet, so install the pipe well below the waterline so it doesn't gurgle, slurp, and burp. If it's below the waterline, it will be invisible too. Making it gentle is a little tougher. Design the intake so the water doesn't create a current going in. One way is to block the end and perforate the sides of the intake end of the pipe. If you don't want to spend your weekends experimenting with different methods, you can hire someone to do the job. See the resources section for leads.

Beavers are animals you can live with in peace. Pacelle says, "We haven't found a case where architecture won't fix the problem."

While homes rarely get flooded, roads often do. Roadside ditches and culverts are perfect beaver dam sites—easily dammed and plenty of water to back up—so they often choose them. There's a wire device, called Beaver Stop, that prevents beaver from damming up pipes; it's sold by D.C.P. Consulting, and it nearly always works. (See the resources section.)

You could destroy the dam, but that would only give the beavers something to do in their spare time. Beavers *like* to build dams, so you're only giving them a reason to do what they enjoy. In theory, it's possible to harass the animal into moving elsewhere, but I can't say who will give in first—you or the beaver. Tearing out a beaver dam daily is hard on a body. Maybe they'll get the hint and look for a different place for their dam and lodge; maybe they'll stay put. If beavers are common in the watershed, they'll probably return to the site.

Since beavers were nearly wiped out in the last century and are only now rebounding, it's illegal to shoot or trap to kill the animals in many states. Hunters agree that it's difficult to shoot a beaver, and traps hold a slow, painful death in most cases. Besides, those types of lethal solutions are only temporary. Since beavers live in groups, you would have to kill the entire family; and then another opportunistic beaver colony would take over the site. It's better to live with the devil you know.

Live trapping is possible but difficult. First of all, the trap itself weighs twenty-five pounds, and you will be filling that with thirty to fifty pounds of beaver. Then you're probably trapping in wetlands or along the water's edge, and you have to set up the trap just so, bait it, and check it regularly. It's work, but it's worth it, says licensed beaver trapper Sherri Tippie.

Six years ago, Tippie had never imagined she could trap a beaver herself. By trade she's a hairdresser. "One day I was scrubbing the floor and watching TV when I saw a story about some beavers they were going to trap and kill on a golf course in Aurora. I thought, 'They don't need to do that.' " So Tippie called the Rocky Mountain National Park and spoke with a ranger there. He said the park would take some beaver. Then she called the wildlife office in Denver. The officer there was incredulous. "Do you know what it will cost to relocate those animals?" he asked. She said she'd do it, and got in touch with delighted officials in Aurora. Then she called a trapper she knew; he told her that no one live-traps beaver—you have to kill them. Never mind, Tippie said, she'd do it without him. So she went to the Denver wildlife division and picked up two traps. They were big and heavy. "I got them home, sat down and started crying. I didn't know what to do." But then she did what we all would do in the same situation; she read the instructions. The first night she went out she caught two beavers.

Since then, Tippie has caught and relocated 110 beavers within Colorado. For bait, Tippie uses apples, castor, and willow or cottonwood limbs woven into the end of the trap. The key to success is firmly staking the trap down. Otherwise, when the beaver swims up to take the bait that's waiting on dry land, the trap flips into the water as it closes behind the beaver and the beaver drowns. "The only trap I recommend is the Hancock trap," she says.

What do you do with beavers? Tippie has had few problems relocating them. Federal, state, and private officials are eager beaver takers. "Beavers do more work for the environment than any other animals," she says. Beavers do good work restoring damaged habitats like former cattle-grazing areas. And trout fishing clubs are interested in keeping beaver because they create wonderful habitats for the fish. Also, the collected silt from a dam and the rotted organic material make terrific soil. "Most of the beautiful meadows in the mountains are old beaver ponds," she says.

If you can catch the critter without harming it, then you have the responsibility of trying to relocate it (or the group of them) to a new suitable territory without any beaver tenants. The worst time to transfer animals, says the Fund for Animals, is in the late fall. The animals just won't have the energy to build a new dam and lodge as well as store food for the winter. They'll starve to death over the winter. Tippie gets around this problem by only trapping in June, July, and

August, when the kits are old enough to be moved. To reduce damage to trees over the fall and winter, she leaves tree cuttings near beaver slides so they'll leave the trees alone. She takes cuttings off the hands of lawn-care companies saving them a trip to the landfill. However, Tippie says, supplementary feeding isn't a good long-term solution, because it's tough work feeding a bunch of hungry beavers. And don't let anyone tell you trapping beaver isn't dangerous doings. Tippie's car has been vandalized by people who misinterpret the good intentions of her work.

Tippie sets great store in sterilization for beavers, because sometimes property owners are willing to have two but not six beaver. She's assisted a veterinarian with eight sterilizations, tubal ligations, or neutering, an expensive and dangerous procedure. She thinks that soon there will be an implant that works too. The adult beavers stay around to protect the dam from other beavers, but they don't have any more offspring.

If you worry about things in general, here's one reason to be concerned about beavers: They are host to many internal parasites, including *Giardia*, an organism that causes "beaver fever" or giardia, characterized by intestinal distress and diarrhea, in humans. Fecal contamination from beaver will render the area below the beaver dam dangerous for drinking and possibly risky for swimming. In the beaver's defense, the parasite is spread by other animals (domestic and wild) and humans as well. Still, if you spot beavers in your vicinity, treat the water before drinking it.

### ❖ BAFFLING BEAVER

1. Protect your trees from gnawing with spray-on products and physical barriers.

2. Install devices to regulate the water flow from a beaver pond.

3. Don't plant expensive trees around a beaver pond; let nature take its course and put your prized trees elsewhere.

4. If you cannot live with your beavers, find a way to relocate them.

# ZEBRA MUSSELS

Wait a second. What are zebra mussels doing in this chapter? Why not in the chapter on bugs or someplace else?

Perhaps the more basic question should be asked: Why is this section in the book at all?

Let me quickly answer those questions: I put zebra mussels in the chapter with beavers because these mussels affect people who live near bodies of fresh water. That's where beavers live.

Second, if you have zebra mussels they are a big problem. If you don't have a zebra mussel problem, be thankful.

The tiny zebra mussel doesn't have much of a brain, so it shouldn't be tough to outwit. Yet zebra mussels have managed to cross the Atlantic Ocean and spread throughout the Great Lakes Basin. The mussels originally came from the Black, Caspian, and Arel Seas, and spread through Europe via the canals in use during the 1700s and 1800s. Sometime in 1986, zebra mussels were sucked up into a ship with freshwater ballast, transported across the Atlantic, and released into Lake St. Clair. And then they started hitchhiking around the Great Lakes.

The average female will produce thirty or forty thousand eggs each breeding season, and the eggs hatch into free-swimming larva, or veliger. Veligers can swim about for nearly three weeks, and then they settle down to the business of attaching to a suitable location. The mussels like to be in flowing water and often settle where they find currents: water intake pipes, canals, and waterlines. Their colonies decrease water flow, and dead and decaying mussels give the water a horrid flavor.

Susan Grace Moore, librarian with the Zebra Mussel Information Clearinghouse explained how thickly they can infest waters, "Someone in the Great Lakes complained that a buoy was missing. They expected that it had just broken loose and drifted off. They sent a diver down to where the anchor would be, and they found out that there were so many zebra mussels on the buoy that it had sunk ten feet below the surface."

The mussels can cause severe problems with boats. The veliger settle on the boat hulls, increasing the drag and decreasing the fuel efficiency. But they can also get sucked into the water intake of the engine cooling system on the boat. If you turn the boat off and leave it for a few weeks and you've got veligers in your cooling system, they can settle out and start to grow and create problems—blocking water, causing engine overheating.

There's documented evidence that they'll settle on other native mussels and crayfish, and they can settle thickly enough that they'll smother the animal.

Homeowners can scrape the mussels from shores, docks, and rocks but must be sure to place the debris in the garbage. Boaters pulling out of an infested lake should drain all water, bay buckets, live wells, bilges, ballast tanks, and even the niches of the trailer. Drain them and let them dry. Adult zebra mussels have been known to survive in cooled-down, out-of-the-water places for up to two weeks.

Hot water (150 degrees Fahrenheit) can kill veligers and adults, and salt solutions have been effective. Antifouling coatings have been used on boat hulls, but they are relatively expensive and must be reapplied often. Some compounds are banned in the Great Lakes and may effect other species, so check with your local extension office or Coast Guard before you use one.

The two-inch mussels don't seem to have any significant biological control here in the United States, and we can't find any use for them. Moore says, "Someone who has steamed them has said they smell like a cross between a dead body and fetid gym socks. They're also quite small so the energy invested in eating them isn't worth the meat return."

The zebra mussel resembles a native salt-water mussel, and only an expert with a microscope can tell difference. It's highly unlikely that you'll find zebra mussels in salt water according to Moore.

## ❖ ZEROING IN ON ZEBRA MUSSELS

1. Drain all water from your boat and trailer. Flush parts with hot water to kill the veligers.

2. Coat your boat's hull with an antifouling substance.

# BUGS, SLUGS, AND SCORPIONS

We all know about ants, hornets, flies, beetles, mosquitoes, cockroaches, and moths: ugly, destructive, annoying. And those flies that attack glass doors in the sunlight with their relentless *buzz buzz buzz*, like a sawmill at full steam.

We all know about the bad bugs.

But even the daintiest of insects can be undesirable company. When my friend Mitch Schultz was walking in the park a Monarch butterfly swooped down at his face, as if out of nowhere. Mitch reacted by swatting at the thing—who knew what it was?!—and sent his eyeglasses on a spiral course toward the nearby bushes. "Ladybug, ladybug . . ." is an endearing children's chant, but have you ever had one of those insects land on the back of your neck and scurry down your shirt?

Theoretically the world would be a better place without bugs. Of course, if we destroyed all insects, we would disrupt the global ecology and probably eradicate our own species in the process, but so what? The world would be free of bugs.

There are more insects in the world than just about anything else. For every person there are 200 million bugs—that's $6 \times 10^{17}$. Insects comprise eighty-five percent of all animal species. From an evolutionary perspective, insects are the most successful creatures on the planet.

My dark feelings about bugs stem from my experiences with the pest species—the cockroaches, mosquitoes, or flies. But I often have to remind myself that over half of all insects are predators or parasites of other insects. So the backyard fogger that kills the wasps also kills the predators of the wasp—not to mention what it does to humans!

But short of killing off all insects, how do we solve what I like to call *the local bug problem?* You know what I mean: ants at the picnic, a mosquito on your arm, yellow jackets at the barbecue, cockroaches who've taken over the kitchen, and the proverbial fly who's landed in your soup? The answer is that each species needs to be dealt with individually. The solution for cockroaches (and yes, there is a solution) won't work for flies. And so forth.

Still, there are four universal truths when it comes to insects, truths that are as immutable as the insects are pervasive. They are:

1. Bugs will eat.
2. Food that is not hermetically sealed will eventually be found by bugs.
3. You can't eradicate every single insect.
4. Pesticides eventually backfire.

Let's spend a moment looking at each of these axioms.

Bugs will eat. In other words, one has to have a certain fatalistic attitude toward insects. Whatever you are growing in your garden or yard, whatever you've got stored in your pantry, whatever you've left outside, bugs will find it and eat a piece of it. Expect bugs: They are like an uninvited cousin who's come to stay more or less permanently, who eats ten percent of everything that's in your refrigerator, and who never gets the hint to do some shopping on his own.

Bugs have skills that even Houdini couldn't match. I don't know what it is, but anything that isn't hermetically sealed will be found by insects. That means the entire outdoors. That means the box of cereal you've had sitting in your pantry since the previous decade. That means the sweaters in the plastic bag. Consider the following items hermetically sealed: refrigerators, Tupperware, pickling jars, and items in Ziploc resealable bags. I don't think I've missed anything. The rest of the universe belongs to insects.

You can't eradicate every single insect. There will be slugs in your garden, beetles in your cucumbers, ants in the grass, flies in your house, and strange, ugly things in your basement. But deal with them when you see them. You can't anticipate what your next bug encounter will be. Sure, you can take steps to discourage insects, but you can't build a glass box to keep them out.

Pesticides eventually backfire. In the short term, they can be effective, but over months or years, the insects become immune to the insecticides. You see, insects reproduce quickly and in such great numbers. Often a mutation occurs that offers resistance to the insecticide, and those are the insects that live to reproduce and pass on the insecticide resistance.

But it's not as if scientists aren't always working on new ones (great job security in that field). So there's always a new insecticide to kill the super insects.

Pesticides have many drawbacks. *They are all dangerous.* They're expensive. They're hard to apply uniformly. In your house, the spray insecticides expose you to lots more toxins than the solid bait stations or even the crack and crevice foams. But even the professionally-applied pesticides evaporate to some extent, so even if you don't touch them or remain in the house as they're applied, you're affected. When applied absolutely correctly, by professionals, the malevolence of pesticides may be limited, but it is still present. Pesticides are especially dangerous when used in the vicinity of pregnant or nursing women, infants, small children, cats, and dogs (the last three will play and fall on lawns and think nothing of eating grass).

In your yard, the spray you apply to your roses will invisibly and insidiously make its way into your neighbor's yard and even into the water table. Or it can do even worse. In the early 1980s, a vacationing man golfed for several consecutive days at a country club in Arlington, Virginia. He came down with a headache, vomiting, rash, and high fever and entered the hospital, where he died from exposure to a fungicide applied to the grass at the golf course. A recent study found dogs had an increased incidence of certain cancers from exposure to lawn-care pesticides.

The National Coalition Against the Misuse of Pesticides, a Washington, D.C., group, reports that little is known about the nature, frequency, amount, or extent of pesticide exposure to the more than

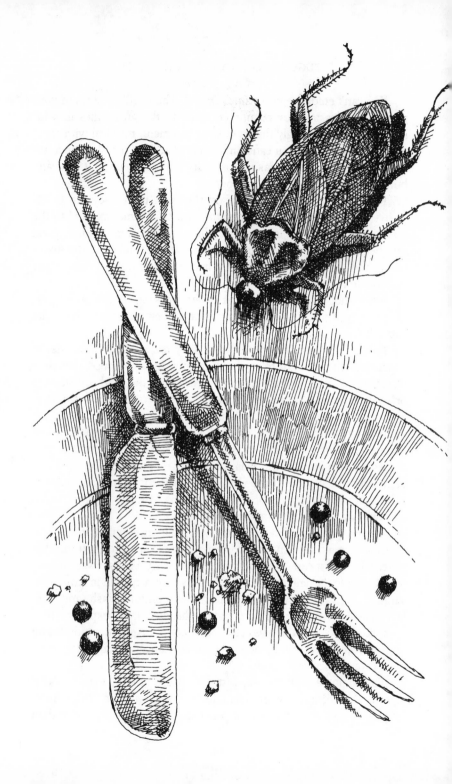

sixty-five million pounds of nonagricultural pesticides applied in and around homes and gardens each year. More than 250,000 Americans become ill from home pesticide use each year. If you ever read a pesticide label, you'll notice that often ninety-nine percent of the pesticide is "inert ingredients." What are they? What effects do they have? "Inert" means that the ingredient doesn't have an effect on bugs—but on humans? There's no such thing as a safe pesticide. Fortunately, there is almost always an alternative to using them.

# Indoor Pests

## ROACHES

When people think of bugs, they think of cockroaches first.

Roaches have been plaguing humans since man dragged home his first mastodon leg and left a greasy campfire ring. They've been around for about three million years and have a biological system that adjusts to whatever pressures they encounter. There are some two thousand different species of cockroach, but only four make unwanted dinner guests: the German, Oriental, brown-banded, and American cockroaches. The smallest is the brown-banded at about a half inch, and the largest is the American (or waterbug) at an inch and a half. Most cockroaches only reproduce offspring once in their lives (whew!), but the German cockroach does it three times. Cockroaches produce 128 to 300 eggs at a time, which are kept safe in a leathery eggcase. The lifespan of the cockroach ranges from six to fifteen months, depending on the species.

There is no place on the face of the planet that roaches cannot inhabit. (Obviously, the outdoor portions of the Arctic and Antarctic are exceptions, but the indoor structures that humans have created there are not.) And we're making things progressively better for them. "As we have gone from living in a Gilligan's Island sort of hut and have evolved to relying so much on structures—not just for where we live and work, but where we store everything we produce—we are creating an ideal environment for arthropods. We reduce air flow, we create dead air spots, and it's perfect for them," says Richard Brenner of the U.S. Department of Agriculture's Agricultural Research Service in Gainesville, Florida. He should know about cockroaches; he constructed a model house, mined it with sensors and probes, and filled it with cockroaches. Now he watches it to see where the bugs go.

Roaches travel quickly and easily, and seem to have a knack for getting into unlikely places. They can squeeze their bodies into slits no wider than a dime. The Edmonton Oilers hockey team once found a cockroach in a tank of purified water at their stadium. Roaches are thigmotactic, or contact-living. They prefer tight cracks and crevices filled with others of their kind. In fact, they won't even mate unless their antennae and feet are in contact with something.

They walk, they crawl, and there's a variety from Florida that can fly (and is making its way North). They are transported in moving boxes, in your morning newspaper, and in the supermarket bags you bring home. So when you move into a pristine, new house—certified roach-free—a few cockroaches will no doubt have hitchhiked along.

And, as most people have discovered, roaches are nearly indestructible. However, that doesn't stop us from trying. And happily, because they're so disgusting, killing them yields no guilt.

The solution I like best, and the one that seems to be most effective, is to buy a lizard that eats roaches. Sure, there's a problem with that solution—a lizard is now running around your apartment. But, frankly, there's no comparison, and if you choose your lizards correctly, you'll never see the lizard.

The lizard of choice is a gecko. I first read about this idea some ten years ago in the New York Times. My roommate and I, who had a typical urban roach problem, bought a gecko, whom we named Spot. From time to time we'd spot Spot on our ceiling, but for the most part, Spot didn't care for us. All he wanted was to eat roaches. As many as he could find.

Geckos are nocturnal, so they tend to do most of their eating at night, when roaches happen to be most active. They like to hide behind refrigerators during the daytime. There are about forty different species of geckos, and they range in size from several inches to eighteen inches long. They don't make much noise, save for a chomping-on-roaches noise and an occasional bark. Many cost less than twenty dollars (some are expensive, but then you wouldn't want hundred-dollar geckos running around—maybe out the door, would you?) and will last a long time as long as you provide water and bugs.

Running out of roaches is never really a problem—at least in most large urban areas—because no gecko (or anything else for that matter) can completely eradicate them. But the geckos will keep the roach population low enough so that you'll possibly never see another roach.

On the down side, you may have a tough time keeping geckos alive in the cold North. And think of them as less of a bug terminator and more of a pet. Geckos need love and care. They'll thrive in the humid, hot South, but are prone to sickness if they're loose in a northern house. You could certainly have more than one gecko running around, but it's bad form to use geckos and poison simultaneously to control roaches. Poisoned roaches could poison your gecko. It's a good idea to capture and observe your gecko from time to time and to try to feed him a few crickets to make sure he's getting enough to eat. Although you can't litter train the animals, you'll find gecko scat dries quickly and can be vacuumed.

If you're not inclined to have a lizard roam your apartment, never fear, new research is always bringing forth data that may help you. The latest research, conducted by Rick Brenner and other scientists at the insect research laboratory in Florida, revealed that roaches detest breezes. They prefer the stagnant air and humidity found in dead air spaces of our dwellings. In their research, air vents were installed on one side of an attic, while the other side remained unvented. After two weeks, the roaches congregated on the unvented side.

Apparently breezes dessicate roaches, whose survival depends on staying moist. What this research suggests is that you should keep your walls and cabinets well ventilated. And if you live in a roach-prone zone and you're renovating a house or apartment, by all means, make ventilation of the attics, wall spaces, and cabinets a priority. Why are cabinets built flush against the wall, anyway? It's not as if doors keep insects—or even rodents—out. Indeed, keeping your cabinet doors *open* may reduce the bug population.

As a matter of fact, Brenner is working with architects to design a home that's less hospitable to bugs. He says, "What we want to do down the road is develop insect-resistant construction practices." Already his lab has produced a ventilation device and a new bait pesticide that effectively get rid of roaches.

Ridge vents placed along the top of your roof work hand in hand with soffit vents under the eaves. When air moves along the ridge vent, it creates a negative pressure under the eaves, where air is pulled into the attic. The soffit vents are screened and hinged, so you can open them and place bait in the eaves to kill attic roaches. You can buy the ridge vents, but the soffit vents and bait will be a little longer getting on the market. (See the resources section for a supplier, page 231.)

There's a strategy to controlling any unwanted pest, and Rick Brenner's makes good sense, the "CIA" approach to insect separation. "In any long-term suppression program, and it doesn't really matter if it's insects or rodents or pathogens, your chances for long-term success are greatest when your efforts are directed against the stages that are concentrated, immobile, and accessible: CIA," he says.

So in developing a strategy to combat roaches, you need to examine the CIA factors, the same way Brenner does his biological research. Where are cockroaches concentrated? How mobile are they? Are they accessible? Can you get to the population that's causing the problem?

One method for assessing your cockroach problem is the old roach motel, a sticky trap that immobilizes the bugs when they step into it. Sticky traps will let you know where you have roaches. Place them along walls, where the secretive roaches roam, and check them daily, keeping a log of the numbers captured. Sticky traps are good for monitoring but not controlling, so use them to monitor the roach population and change them every month. Every time you change them, note how many roaches you catch in a twenty-four-hour period. You can also roam around the house at night, quietly and gently, with a flashlight. Try to spotlight roaches on the kitchen counters or bathroom sink.

Monitoring will give you a good idea of your problem areas. See what you can do to reduce the available food, water, and habitat for the roaches as the first step in your battle against roaches.

Once you know where they congregate, force them out, away from their food and water sources. Use caulk or paint to fill in cracks. You may have to change some design. An elementary school in California couldn't get a handle on their brown-banded cockroach problem. Then they figured out the roaches were leaving their eggcases, safe and warm, in the plentiful pegholes drilled for adjustable shelves in closets. After changing the design, officials had removed one of the essentials for an ever-expanding population—a secure nursery. (School officials also discovered the roaches were feasting on starch-based watercolors and the cornstarch in an indoor sandbox.)

Of course, any treatment for cockroaches begins with better housekeeping. Roaches hate clean apartments. The cleaner your place is, the more likely they are to seek out your neighbor's residence. Don't leave dishes unwashed. If you need to soak a pan overnight, make sure that the entire pan is covered with soapy water. Roaches can swim, but soapy water increases the likelihood of drowning. And the newspapers—throw them out. Your collection of department store bags? Out! All clutter is prime roach habitat.

The second step is to keep all food tightly sealed. Invest in large tight-sealing jars for bulk goods like flour, cereal, sugar, and coffee. Close those lids! Cockroaches will eat cardboard, glue, and anything organic in a pinch.

Keep explorer cockroaches outside your house by tightening up security. Make sure doors and windows are tight-fitting, and seal up holes around pipes and wires that enter the house. Caulk, caulk, caulk. Inspect that box of goodies your kid brings home from school; could there be stowaways aboard?

Finally, remove any sources of water. Roaches need to drink as much as they need to eat, and a leaking pipe will guarantee a fresh supply of roaches forever and ever. Insulate pipes that collect condensation. And don't let water collect in the saucers beneath your house plants.

Sanitation alone won't get rid of roaches. If they are denied food, they cannibalize. One of Rick Brenner's graduate students was able to keep a captive group of cockroaches alive for two years, although he never fed them once.

When you want to kill the pests, boric acid is an old and still effective remedy, though it's slow and may take a month or two to work on an established population. Dr. Donald G. Cochran, entomologist with Virginia Polytechnic Institute and State University says the powder is more effective than the solid bait. Boric acid is available at most groceries or drugstores and is deadly to roaches. Its acute toxicity is low. The loose boric acid is a stomach poison, and roaches die of starvation and dehydration. After walking through the boric acid dust, roaches will ingest the substance when they lick themselves during grooming; it then takes three to ten days for the roach to die. There is no recorded resistance to boric acid.

Sprinkle boric acid lightly (dust it, really) along the cracks and crevices in your kitchen, bathroom, and wherever roaches roam—just don't let it clump up. Use a dust bulb to get the powder into cracks and crevices. If you can, dust the boric acid into crevices and then seal them with caulk. You can also order boric acid bait stations, good for use in damp locations and around children and pets.

Boric acid, a naturally occurring mineral, does not evaporate like many pesticides, and it's not absorbed through the skin. Unlike some pesticides, boric acid will not accumulate in your body. For effectiveness as a pesticide, you need to use the stuff that's ninety-nine percent pure (look for the EPA registration number on the label) so you also avoid the mysterious inert ingredients found in other pesticides.

Some folks think boric acid will be more effective if made into a bait with shortening and formed into balls. They're wrong. Besides, the balls are more attractive to children and pets. It takes time to apply boric acid correctly, but the effects last months, and some people even think they last years.

Many sing the praises of boric acid. A Virginia Polytechnic Institute study found the more toxic pesticides to be sixty percent effective while boric acid had a ninety percent efficacy rating. An experiment at the Alabama Agricultural Experiment Station found that the professionally applied pesticide formula Tempo reduced a German cockroach population by a little over sixty-two percent after a week. But after twelve weeks, the cockroach population was back to its pretreatment levels.

A newly introduced compound, Drion, which contains silica aerogel, causes roaches to die by drying them out. Both boric acid and Drion must still be used carefully, however; small children and pets can also get the stuff into their systems because the toxic compound is exposed to the environment. A teaspoon of boric acid may kill a child. Also be careful not to inhale boric acid as you apply it; wear a mask, gloves, long sleeves, and long pants.

While some of the commercial bait stations contain a growth regulator, you can purchase one on your own. Gencor 9% prevents roaches from reproducing. The spray is applied around areas you see the roaches, and lasts up to six months. Gencor 9% has a good reputation for effectiveness. (See the resources section for suppliers, page 231.)

If all commercial pesticides are unthinkable to you, consider a home brew. Fill a wide-rimmed jar with beer, honey, and banana (or banana peel). Wrap the outside of the jar with masking tape so that the roaches can climb in. Coat the inside rim with petroleum jelly so they can't climb out. By the next morning you'll have more roaches than you could possibly want—or ever knew were there.

Still, Cochran recommends bait traps, like Combat or one of the Raid traps. "They're in a contained unit so pets and kids can't get to them. And you control the dose yourself," he says. These poisons kill in three to five days, and the package recommends the best spots for placement. Target the inconspicuous places where roaches like to travel—behind appliances and in the space near the joint between the floor and wall. You may be able to push them into the wall space through the hole around the drain pipe of your kitchen sink.

If you have a roach infestation, you might be inclined to use a drastic remedy. How do you know if there's a roach infestation? If you enter your kitchen or bathroom during the night and turn on the lights and see roaches on every surface, that's heavy. You'll see scores, not one or two. Also, Cochran says seeing the nocturnal German roach, the most common household pest, during the day is a sign of an infestation. I'd say that catching forty cockroaches in a sticky trap in a twenty-four-hour period is an infestation, too.

Skip the indoor foggers and those automatic bug sprays. The toxicant gets everywhere and on everything. That means on the furniture where you sit, on your bed linens, on the dishes, and on your clothes.

If you're wondering why you have to do anything about roaches—after all, there's no apparent harm—consider this: Cochran points out that there's good evidence that roaches are capable of transferring bacteriological diseases. For instance, your average house roach is not averse to feeding on the soiled litter in the cat's litterbox and then tasting a bit of the cake you left under plastic wrap in the kitchen.

In addition, research in the Gainesville research laboratory indicates that allergies caused by cockroaches are the second most common allergy among asthmatics, affecting ten to fifteen million Americans.

On the subject of what to avoid, add ultrasound devices to the list.

These devices emit high pitches we can't hear, but other animals can. Lowell Robertson, president of Sonic Technology, a California manufacturer of the Pest Chaser ultrasonic device says, "We did a lot of testing with cockroaches, and we found out that you can move cockroaches around and you can change their patterns, which consumers perceive in a positive manner. But you don't really get rid of the cockroaches."

What's going on? Let's say you go into your kitchen at one in the morning and see cockroaches running willy-nilly across the floor, so you plug in an ultrasonic device. The next night, when you go out there and turn on the light at the same time, chances are, you won't see any cockroaches running around. Explains Robertson: "An average consumer's perception is, 'This is terrific, it got rid of the cockroaches.' Well, no it didn't." In fact, he says, the roaches are still there; the sound just forced them to vary their routine—to take up residence in what he calls "sound-shadowed areas."

At this point in the discussion of roaches, it's worth pointing out that not all people are interested in getting rid of roaches. Hard to believe, but true. There are at least two groups of people that like cockroaches. First, and most obviously, are exterminators, whose living depends on these insects. Second are professional and amateur cockroach racers. In 1990, the annual Great American Bug Race, held in West Palm Beach, Florida, had sixty roaches lined up antenna to antenna. It's a tough contest, the Great American Bug Race, as some roaches didn't know what to do and dashed into the spectators. In 1990 the winner received $150. (Pays for a year's supply of Raid.)

# FLIES

Face it, flies will get in the house from time to time, and they're annoying. Generally you can do them in with a flyswatter or a rolled-up magazine. Flypaper is unattractive, but it works and doesn't harm people living in the house like an indoor spray. Today you can even buy sticky flypaper that's hidden inside a decorative box so you don't have to look at fly cadavers all day.

Flies harbor disease. They're pretty common in restaurants. If you see a lot of flies in a restaurant, it's a good guess that the establishment won't get four stars when the health department comes around. If you see absolutely no flies in a restaurant, it's a good guess that the place uses a lot of pesticide. Take your pick.

When I went to summer camp each bunk had a pesticide-infused "no-pest strip" hung in the middle. What a bad idea: This no-pest strip killed flies all right, but its presence also meant that we were inhaling pesticide vapors all night long. Avoid all aerosol pesticides because they will get into your lungs.

New England residents have to deal with cluster flies, which hibernate in the walls of houses and emerge inside warm rooms on sunny days. Carl Taylor of Vermont contributes to a benign (for the environment, not flies) control method. He keeps a Venus fly trap on the kitchen windowsill. While he admits that his children actually have to catch the flies for the plant, it does dispose of two or three flies a week. At any rate, it gives the kids something to do on long winter evenings after the homework is done. Other New Englanders merely vacuum the pests up and toss them outside into the frigid winter air where they die.

In the great outdoors, traps can help reduce the number of other flies in an area. Peaceful Gardens, an organic garden supplier in California, sells a fly trap that attracts the insects with smell, a strong ammonia odor. This trap is designed to be used around animals, say, in a horse barn, and it's so smelly that you can't put them by the front door of a house. But they would work away from your house.

The flies are attracted to the ammonia smell of the bait in the trap, fly up into the trap, and can't get out. Says Mark Fenton of Peaceful Gardens, "After you get a couple of thousand flies in the trap, the buzzing itself attracts others from a long distance."

## ANTS

Getting rid of ants is no picnic. Ants seem to view the individual ant as expendable, the survival of the species as all. So, smash one with your kitchen sponge, and instead of scurrying away, they send out the reinforcements: Another 250 ant soldiers march inexorably under the door, up the cabinets, across the wall, and down into the sink as you stand with your sponge poised in despair.

Nothing seems to work. Clean the sink, turn your back, and next thing you know, they'll have formed a line into the garbage. Take out the garbage, and you'll find them swarming over that single Cheerio under the kitchen table. Scour the kitchen, and they show up in the bathroom.

When our daughter, Karen, drops a cracker on the porch, which happens every time she's on the porch, if we don't pick it up right away, the ants carry it off. Without fail.

Their strategy, basically, is to use their numbers to wear you down, so you'll surrender and declare your home a national indoor picnic site. But do not despair, for effective strategies await the courageous.

There are hundreds of species of ants, but the only one that commonly takes up residence in our homes is the pharaoh ant (also called the sugar ant). It builds nests indoors. Pharaoh ants are small, reddish insects, about a twelfth of an inch long. They live in colonies with up to several hundred queens, and worker ants take care of everything.

About the worst thing you can do against pharaoh ants is to use a conventional pesticide, which makes the colony split up into multiple colonies. Because they have so many queens, they can make lots of colonies—all over your house. Boric acid works best against them. (See the section on cockroaches for more information on boric acid.)

Ants aren't going to walk through it and then lick it off like the cockroaches do, you have to slip them a mickey—put it in their food. The worker ants eat the poison bait, and then take some back to the nest to feed the queen and young. The liquid forms of boric acid, marketed under the name Drax are more effective for ants, and you can buy the bait from a number of mail-order catalogs. Or you can make your own.

Mint-apple jelly is one of the more tempting baits for pharaoh ants. Mix one teaspoon of ninety-nine percent pure boric acid into one-third cup of the jelly. Place the bait along a piece of tape, in bottle caps, or just on the floor or counter where you have seen ant activity. Don't block their trails with bait. Each daub treats about twenty-five feet. Check the bait (commercial or homemade) every three to five days, and replace the dried-out bait. (If you're frugal, refresh the bait with water.) You can also use corn syrup for bait.

The hard part about this treatment is watching ants feed at the bait stations while you stand by and do nothing. They'll crawl away and die. It takes one or two months to clear away the ants using boric acid.

Other ants you see around the house are visitors from outside. They're just visiting your food supplies, and they keep their nests out in your yard. With these fellows, you need to find where they get in and block the exit. If you're inclined, you can kill them with a poison, but not a bait, which would just encourage more deadbeat ants to come on in and eat your food. Try sprinkling boric acid powder (the ninety-nine-percent pure stuff) around the areas of activity. A product called pyrethrum made from pyrethrum flowers, a member of the chrysanthemum family, works well. Just because it's a botanical, doesn't mean it isn't deadly; it's twenty-five percent more acutely toxic than boric acid. Dust the areas where you find ants. Pyrethrum kills a variety of household pests and breaks down well. (However, clearing land for cultivation of pyrethrum is destroying mountain gorilla habitats in Rwanda, near the camp Dian Fossey established to study the species.)

We had an ant problem in our house. My wife attributed the problem to crumbs I left behind when I sliced bread, but I thought the infestation was more systemic. I agreed, however, to be tidier to see if that helped alleviate the problem.

Meanwhile, ants were still appearing in through the cracks in the wood floor in our dining room. We tried several strategies including filling the cracks with pulverized tobacco, ultravacuuming the dining room, and keeping extra-bright lights on. (Don't ask me why we thought bright lights would work; we're experimenters.) Nothing we tried from the inside would help. So we figured we would deal with the problem at the source. under the dining room.

Fortunately, our dining room is raised head-height above the ground, so it's possible to walk beneath it. *Unfortunately*, the only outdoor solution we could think of involved pesticide. As you probably have discovered, I'm not a fan of pesticide—especially in my own home—but here we had no alternative. So I bought a selective, ant-and-roach-only pesticide. (Selective pesticides are often less toxic than broad-spectrum pesticides, designed to kill everything.) Peggy volunteered to spray, and, after donning rubber gloves, she went to her task, spraying the sites where ants were living and entering our house. Only those particular areas needed to be targeted. After spraying, Peggy washed her clothes and took a shower—another good practice for using pesticides. The ants haven't come back.

Another possibility is household cleaner. Interestingly, household cleaners make a terrific spray that's effective against ants as well as other bugs. Often, household cleaners work faster than pesticides. Windex, Formula 409, just about any spray cleaner, will do the trick when directly sprayed on the ants or their habitat. Kills 'em dead.

Conveniently, after you've murdered the bug, you take a cloth and wipe that part of your house clean—think about that: Accomplishing two important tasks at the same time! Even less toxic than aerosol sprays, but just as effective against ants, is a mixture of water and soap. Next time you're done with a bottle of Windex, fill it with soapy water and label the bottle *Bug Spray*. Your homemade mix or some commercial insecticidal soap works well but is harmless to humans.

While cleaners work on the ants that you see today, the best strategy against the visiting ants is to block their access. So spare a few and watch them to see where they exit; they generally follow chemical trails they have left in and out of the house. A caulk gun works well to block the holes. They're persistent insects, so don't throw away the caulk gun after the first time; you'll use it again. You should also take a look around with an eye to cleaning up food sources. Small ants can climb up loosely closed screw-top jars to die sweet deaths in honey or liqueur. They're attracted to open grains, pet food, spills, and garbage.

Ants in the yard are rarely a problem, but if you notice a population explosion, look for food sources: fat spilled under your grill, exposed garbage, partially burned refuse, a compost heap that's not composting right, or pet food. Sometimes ants may be attracted to feed on the sweet "honeydew" produced by scales and aphids that are attacking plants in your garden. Deal with the aphids and scales, and eliminate the food source.

You may drive out the ants with water after you have eliminated the food source. Flood each ant hole with water from the hose, pressing the nozzle against the earth, or pour soapy or boiling water into the hole. Pesticides like boric acid or pyrethrum used outside will kill a wide range of insects, so be careful if you use it in open spaces.

Keep in mind that not all ants are bad. There's a beneficial side to ants, as well: They are the sworn enemy of termites and eat fly larvae. So if you have an ant infestation inside your house, only attack that problem. Don't eradicate all the ants in the neighborhood, or you might invite a plague of other insects.

## FLOUR BUGS

I almost became a believer in spontaneous generation when I opened a tightly sealed jar of flour and found it contaminated with wispy strands and weevils. Where did they come from?

The little brown moths you see flitting around your pantry will lay eggs in your dry goods, and soon little weevils will hatch there. They spread from one product to another, and pretty soon everything is infested. So you throw everything out and get new products.

If you freeze your new flour, millet, or other dry goods for a few days before putting them on your shelves, you may prevent a new infestation. The freezing kills the weevils in the food.

If you want constant shelf policing, try some flour bug traps. A new twist on old, familiar flypaper, the pheromone lures, attract Indian meal moths and warehouse beetles (commonly known as flour bugs) that like to attack your flour, grains, dried fruit, dried milk, expensive Dutch cocoa, and other cabinet staples.

These lures, available through a number of organic garden suppliers, are small, rectangular boxes impregnated with sex pheromone—an attractant. The inside of the box is lined with a sticky surface that traps the bugs. The lure works for about twenty weeks before you have to replace it.

## FLEAS

Fleas are sneaky little critters. They hitchhike into your house on animals and then move in. They make a home of your carpet and furniture, reproduce, and plague you and your animals. So if you only treat your animal for fleas, they'll be reinfected from the resident fleas.

You may notice fleas when your pet starts to scratch a lot. They're worst during the warm spring and summer months. Check your pet for fleas, tiny black specs next to the skin. You may also notice red spots (flea bites) and black and white pellets (fecal matter and eggs).

Fleas don't have any wings, but they sure can jump. But you know that if you've ever picked a flea off your pet. Just when you're sure you've smashed the little critter between your finger pads, it jumps away. The best way to kill fleas you pick off your pet is to fling them into a solution of soapy water or alcohol. During flea season, groom your pet regularly with a fine-tooth flea comb (fleas will stick to it if you put a bit of Vaseline at the base of the comb, and the Vaseline won't hurt the animal).

If your pet has more than a few fleas, you'll want to treat its entire body. Avoid the most noxious compounds and stick to insecticidal soaps or shampoo or dip treatments with diatomaceous earth or silica gel in them. Ingredients to avoid: piperonyl butoxide or petroleum distillates.

Botanical insecticides like pyrethrum or rotenone are very effective both on the pet and around the home. There are a variety of pet shampoos that combat fleas yet don't contain harmful chemicals; they contain insecticidal soaps, oils of citronella, cedarwood, and eucalyptus—ingredients often found in natural insect repellents. Herbal flea collars are also impregnated with similar oils. Garlic and nutritional yeast added to your pet's diet may help it repel fleas; you can buy it in a prepared form with additional herbs, minerals, and vitamins added or you can simply add yeast and garlic to the pet's food. Ask your veterinarian for advice on how much to use.

You need to treat the house too, so wash the pet's bedding if you can. There are insecticidal soaps for use around the house. Two growth regulators, hydroprene and precor, prevent fleas from reaching adulthood, and both are effective tools for combating a flea infestation. There's even an electronic flea trap that attracts fleas from a room to a warm area covered with sticky paper.

While ultrasound devices don't work on most insects, they do seem to affect adult fleas. Says Lowell Robertson, president of Sonic

Technology, sound "seems to interfere with fleas' ability to source the blood host. Our field test trials show you can effect a substantial reduction in the flea population." These devices, however, cannot "stop the animals from getting fleas on them when they go outside," he points out. An ultrasonic flea collar won't work because the sounds will get stopped or deflected by the animal's body. Save your money for a new leash or scratching post.

## ❖ INTERDICTING INDOOR INSECTS

1. Make your home inhospitable to bugs. Make food inaccessible. Wash your pet's bedding often. Vacuum often and change the bags.
2. Use insecticides sparingly.
3. Avoid gadgets that seem to be too good to be true; they are.

# Outdoor Pests

Here I'm not going to focus on garden pests, really. In the resources section, I've listed some good sources for garden pest information; the topic deserves an entire book—and has filled scores. I'm more interested in the bugs you'll run into at a barbecue or on the lawn.

## MOSQUITOES

Mosquitoes are one of the most pesky critters around, but many deterrents are distasteful—some are foul-smelling and contain DEET, (N, N-diethyl-meta-toluamide—see why they call it DEET?), a chemical that may be carcinogenic. DEET is in some of the more popular outdoor repellents. Then there are the bug zappers, which are indiscriminate in their effect, and kill all insects. When you're sitting in the backyard listening to the rapid *zip!* of the zapper, you're hearing good insects go up in smoke. The zapper's light also attracts all the bugs in the neighborhood to your yard. If you want to use a zapper, give one as a present to a neighbor a few houses away. Backyard foggers and sprays also kill indiscriminately—and expose you to the pesticide.

What to do? There are several ways to keep the mosquitoes at bay. You could stay indoors, or in a screened porch. But you can't always stay away from the mosquito. Besides, from time to time one mosquito will get under the covers with you, heading straight for your inner ear.

Insect repellents are probably the most widely used weapon against mosquitoes. Repellents based on natural ingredients have recently been introduced to the market. Cal Saulnier of Plow and Hearth Catalog (see the resources section) says, "The natural insect repellent, made with pennyroyal and citronella tested pretty well, and with the controversy surrounding DEET, it can't hurt to try it."

What's wrong with DEET? Who knows? It's in most of the repellents you buy at the drugstore. People have been using it for about thirty years, and most people don't think twice before they spray on the Off! In the summer of 1990, the EPA issued a consumer bulletin warning that a small segment of the population may be sensitive to DEET. Applications may cause headaches, mood changes, confusion, nausea, muscle spasms, convulsions or unconsciousness, and children are especially vulnerable. The agency is still researching DEET's alleged effects: cancer, birth defects, and reproductive problems. Eventually we'll know. Decide for yourself whether you want to be part of the research.

Still, the most effective repellents around contain DEET, although you don't need to use anything near a one hundred percent concentration of DEET for it to be effective. A Rhode Island entomologist, Richard Casagrande, wrote an article on repellents in *Harrowsmith Country Life* magazine. After testing several of the natural repellents, he found Bug-Off, Cedar-al, Natrapel, and Avon's Skin-So-Soft to be most effective—although none of these are as effective as products containing DEET. The natural repellents contain volatile oils, and the effectiveness of the concoctions quickly dissipates. Any repellent works best if sprayed on, not rolled or rubbed on, and be sure to treat your clothing (DEET melts plastic and tastes awful). Be sure never to use any product with more than fourteen percent DEET on children. Never use DEET products on newborns.

Low-tech solutions include citronella-scented or pyrethroid torches and candles and mosquito netting, which is useful in areas—like the north woods in late spring and early summer—where mosquitoes are expected to be especially numerous or aggressive. You can buy mosquito-netting hats from several different outdoor outfitters, but be sure to treat the top of the hat with repellent or mosquitoes will bite the top of your head. Wood lore has it that eating large quantities of garlic can keep mosquitoes away. (Although, as you know, there are other drawbacks to eating garlic.) Some people even use naphthalene granules, a product purported to repel everything from deer to squirrels, and of course moths, since naphthalene is the main ingredient in moth balls.

On the higher-tech end of the spectrum are ultrasonic insect repelling devices. The plug-in device is supposed to mimic the sound of the dragonfly, a natural enemy of mosquitoes. But some entomologists are skeptical, as are some in the outdoor-products industry.

For example, Sonic Technology's Lowell Robertson says tests proved disappointing. "Early on, we did extensive testing on mosquitoes, because mosquitoes are a high visibility insect and one that people would definitely like to get rid of," he says. "We found out ultrasound isn't a very good tool for getting rid of mosquitoes." And Cal Saulnier says his company, Plow and Hearth, stopped carrying an ultrasonic insect repeller, discouraged by the ten percent return from dissatisfied buyers.

Still another tactic is to attack mosquitoes at the source: to cut down their populations by getting rid of their breeding grounds. Mosquitoes aren't picky about nurseries: All they need is a small amount of water—a flower pot saucer, a fish bowl, a garbage can, a birdbath that hasn't been changed for several days, a pool of water behind your garage. Even the smallest pool of water can give mosquitoes a place to breed.

The best time to hunt for stagnant water is after a rainstorm or after you water the lawn. Trash can lids and wheel barrows are places where water can pool. Empty them. Then, try to set some traps. Leave a pool of water around, perhaps in a soup bowl or flower pot—but add soap to the water. Female mosquitoes will land on the water to lay their eggs, but these mothers-to-be will die in the process when they drown in the soapy water. If you have a fish pond, you can even stock it with mosquito larvae–eating fish, of the *Gambusia* genus. The bacterial strain *Bacillus thuringiensis* (Bt), used by organic gardeners against pests, will kill mosquito larvae, which eat the bacteria and die.

Another way to cut down on mosquitoes is to install a bat house in your yard. A single bat can consume up to six hundred insects an hour. (For more on bat houses, see the chapter on bats.)

# TICKS

Even before Lyme disease was in the popular vocabulary, ticks were scary creatures, violating your protective skin to suck the life's blood from your body. Today we know that in addition to their general repulsive behavior, they can transmit the spirochete of Lyme disease as they feed. Lyme disease affects thousands of people each year and has been reported in every state but four. Ticks also spread Rocky Mountain spotted fever, babesiosis, Q fever, and tularemia.

Ticks take a long time to eat their fill of blood (days or weeks), and they feed on more than one type of animal. Some ticks in the larval stages feed on mice or voles and then graduate to larger mammals like dogs and humans when they're older. Ticks were put on the earth to make other insects look good.

On your pets, don't forget to check for ticks in the fall and winter, and don't neglect the ears and feet where the ticks like to hide. Ticks and nymphs, the immature tick, are most active in the spring and summer, though some adults are active through the winter. The brown dog tick, the main dog-nemesis, will move into your home and lurk about in cracks and crevices waiting for a ride and a meal, probably with the dog. Ticks that carry Lyme disease are of the greatest concern.

Once acquired, Lyme disease has no cure (for us or animals), but the disease can be arrested with antibiotics. The disease is carried by the northern deer tick in the East and the western black-legged tick in the West. The nymph, which experts say has the greatest likelihood of transmitting the Lyme disease, is most active in the spring and summer and is tiny, a pin speck of an animal. You should check your pets for ticks twice a week during prime tick nymph season.

Because Lyme disease transmission has a time factor, the sooner you remove the parasite, the lower the chances are of getting the sickness. Use a pair of tweezers to grasp the tick as close to the mouth parts as possible. Pull straight up slowly and steadily until the tick is removed, and then treat the bite with alcohol or iodine. Destroy the tick by drowning it in alcohol or soapy water, but remember that you'll expose yourself to any tick-borne diseases if you smash it between your fingers. Don't leave a dead tick attached and don't cover the tick with toothpaste, Vaseline, or anything else waiting for the animal to smother, you'll only prolong any exposure to disease. If you're willing to risk a burn, using a hot needle will kill the tick.

If you must venture into tick habitat during prime tick season, wear light-colored clothes that cover your legs and arms. Dusting your clothes with botanical powders containing pyrethrum will help too, or try repellents with DEET, citrus oils, or insecticidal soap. Clean up brush and debris around the house that will attract rodents, and consequently tick larvae and nymphs.

## BEES AND WASPS

Say "bee" and most people think of a cute yellow insect on a flower. Or maybe they remember being stung when they walked barefoot through clover. Or the great killer bee skits of the classic "Saturday Night Live" television shows. But say the word to a resident of Scottsdale, Arizona, and you'll probably be hit with a barrage of bee stories. It seems that a warm, wet spring in 1991 led to a bee population explosion in the surrounding desert. Then the water dried up, and the bees went for some of the swankest pools in Scottsdale. One resident reported seeing as many as ten thousand bees at his pool a day! They clogged skimmers and terrified pool owners.

There was really no solution to the problem, and without the pools, the bees probably wouldn't have made it through the summer. People tried lures of kiddie pools. No luck. Now the residents are hoping for a cold, dry spring. But aside from occasional freak occurrences in nature, bees are seldom problems.

Yellow jackets are another story. Most stings come from yellow jackets, the black and yellow wasp that can sting repeatedly and live to sting another day. Sure, there are the big bumblebees in the honeysuckle, but they're slow. Yellow jackets are the terrorists of the wasp world. My wife was inexcusably stung by one while we were attending an outdoor wedding. Had she been stung at the exact moment the minister asked whether anybody objected, the whole affair could have inadvertently been called off.

Yellow jackets are frightening. They're the ones that swarm around your head and try to get in your soft drink. They swarm garbage cans. They love sugary foods and proteins, and people are often stung while dining al fresco or accidentally ingest a yellow jacket with a bite of hamburger. Yellow jackets are just as excited by Diet Coke as they are Classic Coke.

Yellow jackets live in colonies begun each spring by a lone queen. She begins the nest and lays some eggs, and then after about a month, some workers emerge from the larval state to help expand the nest and feed the other larvae. Soon the queen only has to lay eggs and the workers take care of the rest. At the height of the season, new queens and males are produced; they will mate and only the queens will hibernate through the winter. As it gets cooler, activity increases in the nest, and the workers concentrate on quick energy, high-sugar foods instead of the proteins they sought earlier in the summer. They're more aggressive late in the season. The yellow jackets usually live just one summer, then they die off, leaving the queen to hibernate. A cold winter kills a lot of the queens, but after a warm winter, more queens survive and produce many more offspring.

In general bees and wasps only sting when provoked. How do you know what provokes them? For starters, yellow jackets nest in the ground, in old stumps or in logs. If you dig up a nest in your garden, that's provocation. You may be the victim of a mass attack. Yellow jackets are the most unpredictable of the bees and wasps and often sting for no apparent reason. Yellow jackets can sting repeatedly. And often will.

When you're stung by a wasp or bee wash the wound and then reduce the pain by applying ice, meat tenderizer, wet tobacco, or a commercial sting-treatment product to the area. If it's a bee sting, first scrape along the sting with a plastic credit card, your fingernail, or a driver's license to take the stinger out (bees, unlike wasps, leave a venom sac in the wound with the stinger). Signs of allergy include: difficulty in breathing, severe swelling, or hives. Anyone stung in the mouth or throat should visit the doctor: a hive, from a sting in the neck, could block breathing passages. Nearly twenty people in the United States die each year from shock after being stung.

If you can keep your head about you when a yellow jacket is about your head, you may avoid a sting. Don't wildly swat them, but gently brush them off or wait for them to fly away. They're attracted by perfumes we use, including those in sunscreens, shampoos, deodorants, and hair mousses. They're also attracted by brightly colored clothing. Yellow jackets also need water and may be lurking in wet clothing or towels or around the pool, so watch where you sit. Yellow jackets cause the biggest problems when there's food around.

Before officials at Virginia's Great Falls National Park started managing the yellow jacket problem, visitors suffered about a thousand stings a year. Now, instead of serving sweet drinks in open containers, food concessionaires added lids and straws; they use closed garbage containers; and they pick up the trash more frequently. Stings have dropped to forty a year.

If you must eat outdoors, try one of the yellow jacket traps listed in the resources section. You can also attract them to plain old flypaper if you add an attractive bait—dog food, ham, and meat scraps in the early summer and sugary foods like jams, syrups, and fruit later in the season.

Destroying yellow jacket nests is dangerous business, and something best left to professionals who have the proper gear. But if you have skunks or raccoons around your house, you'll be interested in this: A Common Sense Pest Control newsletter reported a way to get someone else to do the dirty work. One evening a naturalist with the East Bay Regional Park System in Oakland, California, poured honey over the entrances to ground wasp nests. The next morning, the wasp nests were all excavated and destroyed. Many species dig up the nests to get the honey in the larval chambers.

# TERMITES

There are two things you need to know about termites:

First, if you don't have termites take steps to maintain this status quo.

Second, if you do have termites, pesticides are not necessarily the only option you have (despite what termite eradication companies may tell you).

There's a vast literature about preventing termite infestations, so I only want to highlight the most salient tactics:

**1.** Keep firewood and other termite foods away from your house. Don't stack the wood against an exterior wall. Remember, termites like any kind of wood: Think of firewood as termite bait.

**2.** Keep your basement and the area outside your basement as dry as possible. Moisture is essential for the survival of termites. This means making sure that the drain spouts work properly and that bushes don't come in contact with your house.

**3.** Keep your house well painted. Cracked and peeling paint offers termites a way in.

But what if you have a termite problem? Nematodes are a sensible alternative to pesticides. These tiny, wormlike creatures are a deadly parasites to termites: Unlike pesticides, nematodes won't harm your children or pets—or anything other than termites.

Finding a company that uses nematodes won't be all that easy, although your local environmental groups may give you a lead. Also, nematodes can be ordered through the mail from a number of organic gardening supply companies.

There are two other potential solutions that you might consider. The first is freezing your house with liquid nitrogen. This lowers the temperature in the walls to twenty degrees below zero—a frosty temperature that spells death for termites. (That's why there are no termites in Alaska.) It may also kill all the roaches and ants. Make sure your pipes have been drained beforehand or the water in them will freeze, expand, and possibly burst the pipes.

Alternatively, heating your house to 120 degrees Fahrenheit by blowing hot air through it will rid your house of termites. And this technique will rapidly kill any other insects that happen to be around. As a precaution, remove all candles before you pursue this option. And cans of Coleman fuel!

More information about the nontoxic control of termites and other insects can be obtained from the National Coalition Against the Misuse of Pesticides.

# BEETLES AND GRUBS

Aging yuppies don't always remember that the four from Liverpool weren't the first. The bug variety of beetle has been around much, much longer. Beetles are hard to control, even with pesticides. This is especially true with Japanese beetles, which feed on over two hundred varieties of plants—most often the ones that are hardest to cultivate. As beetles, they eat your roses; as grubs, they create patches of dying grass as they feed on the roots.

But there are some simple steps you can take to help eradicate them:

1. Avoid beetle traps. Traps use a scent that *attracts* insects. You'll trap a lot, but you'll attract even more. In her *New York Times* column, "The Cultivated Gardener," Anne Raver reported that studies showed setting traps every two hundred feet throughout a neighborhood would reduce the numbers. (People who can't agree on zoning variations for decks won't agree to this!) However, if your land is large enough, it's okay to place a trap in a distant corner of the yard where you may not draw beetles to your prized plants.

2. If you want to try to vacuum Japanese beetles off your plants, or hand pick them, morning is the best time—that is when they are most slothful. Then squash the bugs between your fingers, under your heels, whatever. Just kill them. If you're squeamish, drop them into a container of soapy water.

3. Pay attention to the life cycle of beetles. Japanese beetles lay their eggs in July and August; the less you water your lawn during

those months, the fewer beetles you'll have. The beetles feed until October and November, when they move below the frost line and wait out winter. By March, they're eating again. You can spot grubs if you peel back a layer of sod and peek beneath, and if you see ten or more per square foot, you've got a problem.

**4.** Try using *Bacillus popillae*, aka, milky spore, to kill beetles in their grub stage, when they live in the dirt under your lawn. Milky spore effectively fights soil-living grubs of the Japanese beetle, rose chafer, oriental beetle, and some May and June beetles. It's sold under the name Doom or Grub Attack at garden centers or you can order it from a variety of organic supply catalogs. You apply it by sprinkling the dust onto the lawn and then watering it so it sinks in. Treating for lawn grubs also stops any grub predators like moles or skunks from tearing up your lawn in search of a meal.

The spore kills the grubs and remains effective in the soil for twenty years. Plus, when they die, the decomposing grub bodies release more spores into the soil. It's harmless to humans and pets and won't hurt your earthworms either. Unfortunately the milky spore takes three years to get established, so you won't see the results immediately.

A quicker killer can be found in parasitic nematodes strain Hh, also available from organic gardening suppliers. These parasitic nematodes enter the host body and release bacteria that will kill the grub in forty-eight hours. The nematodes then feed on the grub's remains and reproduce.

The most creative grub killing award goes to some Colorado State University entomologists: They used soil aerator sandals. In the early and late summer when the grubs were feeding heavily near the surface, the entomologists put on their spiky sandals and methodically walked over a lawn several times to get two nail insertions per square inch.

# SLUGS

I planted marigolds next to my tomatoes because marigolds are supposed to discourage pests. Within a few days, they were mowed down.

Gone. I wanted to blame the harmless pillbugs that abound in the soil. But I went out at night with a flashlight and watched. Slugs. Slugs did it.

I set right to work. I buried several jars up to their rim in the soil. Then I filled the jars with beer. (I used Old Milwaukee, no sense in wasting the imported stuff.) The next morning, I had jars full of beer and slugs. I poured the stuff into my compost and set the jars in place again, refilled. Nothing ever touched my marigolds again.

If you want to go high-tech, you can buy the commercial equivalent of a jar of beer with a roof and a removable cup in Garden Sentry. A number of garden supply catalogs sell something like the Garden Sentry (see the resources section for names of suppliers). The device gets buried up to its rim in the garden and is filled with beer or commercial slug bait to lure slugs into it. Once inside, slugs drown, since they can't get back up the steep sides. When you need to get rid of dead slugs or you want to refresh the bait, you simply lift out the sieve-like, removable cup and leave the device in the ground. The roof keeps the bait from getting diluted, but may cause problems for lawn mowers when the device is placed on the lawn.

## OTHER YARD AND GARDEN PESTS

Yet another technique—a favorite of mine—is to attack garden bugs with a hand-held vacuum cleaner. Stalking your garden, swooping bugs into this high-tech (well, relative to bugs it's high-tech) contraption offers a great deal of satisfaction. Besides, it's very effective. Vacuum cleaners are most effective against leaf-eating insects such as aphids and beetles. Empty the bag into soapy water to kill any insects that survived being swooshed into the machine.

Unscented Sticky Traps also work well. Yellow ones seem to attract aphids, whiteflies, leafhoppers, black flies, moths and gnats; white attracts tarnished plant bugs. You can clean the sticky trap's surface with vegetable oil and, to reinvigorate it, brush on a liquid sticky coating like Tangle Trap. In addition to invigorating sticky traps, you can also use the Tangle Trap or a convenient spray or tube form to cover different surfaces.

You can also use beneficial bugs to do your dirty work. Predator bugs include ladybugs, lacewings, praying mantis, spined soldier bugs, and trichogrammas. You can order the insects from several organic gardening suppliers; check the resources section for the names of just a few.

Some other powerful, "organic" weapons against yard pests include nematodes and bacteria. One bacteria, *Bacillus thuringiensis* or Bt, controls a plague of insects: cabbage loopers, cabbageworms, diamondback moths, tomato hornworms, tomato fruitworms, grape leaf folders, and gypsy moths.

## ❖ OUTLAWING OUTDOOR INSECTS

1. Try to attract natural predators to combat bugs.

2. Remove brush and debris that attract the undesirables and gives them a place to hide.

3. Avoid broad-spectrum pesticides that will kill all insects—good and bad—as well as some birds and mammals.

# SCORPIONS

We all know what scorpions look like. We've seen them in James Bond movies, on astrological charts, in zoos, and maybe (eek!) in our houses. They're the scary-looking things with long, curved tails equipped with stingers on the tips. Stingers that can deliver poisonous stings.

Scorpions, which are closely related to spiders and mites, are found throughout most of the world. There are eight hundred delightful species. In the United States, they're most common in the Southwest (most of the species that live there aren't deadly but pack a powerful wallop of a bite). But don't breathe a sigh of relief if you live outside that region. "You find them up north, maybe even in Montana, all the way east, in Florida, all the way through Mexico," says scorpionologist Carl Olson, associate curator of entomology at the University of Arizona.

Scorpions are solitary, nocturnal animals, and only come together for mating. It's hard for us to tell the difference between males and females, but they can tell. After a complex mating dance in which they entwine tails and scoot around, the male fertilizes the female's eggs, which she carries internally. If the male doesn't scoot away, the female may kill and consume him after they mate.

Depending on the species, the young are born after several months to a year. The young rest on the mother's back for days, until they molt and are ready to hunt on their own. Scorpions may live for several years.

Although they can spend long periods of time without food, they prefer not to. They seek animal fare: insects and small invertebrates, mice and other small vertebrates. They wait in hiding and surprise their prey, but they don't chase it down.

They grab the prey with pinching front appendages and sting and paralyze the animal if need be. Eating takes a long time for scorpions because they don't chew their food. Instead, they ooze digestive juices over the victim and then suck up the predigested meal.

In turn, scorpions are eaten by birds, lizards, mammals, insects, spiders, and other scorpions, who disable the stinger before liquefying the prey. Despite popular folklore, scorpions don't commit suicide by stinging themselves.

All scorpions bite, but the bites of most varieties of scorpions are rarely fatal to humans. However, the bites do cause pain and numbness and can bring on local paralysis, fever, and general flu-like symptoms for a few days. The exception is the buthid family, whose bites can kill a person in a few hours, although there is an antidote. You won't find a buthid in your bath, though. These inch-long, beige animals live in the desert, and keep backpackers on their toes because they're known to crawl into sleeping bags and boots.

Most of the time, you'll find scorpions outdoors, in washes, rocky areas, under rocks, in burrows. However, there is one scorpion that likes the indoors more than other members of the species: the bark scorpion.

The bark scorpion, prevalent in the Southwest, gets its name from its preference for old wood. It likes wood piles and rotting walls. It has a special fondness for old, dead palm trees, where it crawls under the bark. This scorpion also likes spending time under rocks.

However, the bark scorpion has also discovered the great indoors. In fact, it almost seems they prefer man-made habitats to those provided by nature. (Of course, those man-made habitats are taking over nature, so scorpions, which have been around some four hundred million years, are just making another adaptation.) Says entomologist Olson, "They're adapting very nicely to the habitat we give them."

How do they get into houses? According to Olson: "There always seem to be little cracks hither and yon. They're not hitchhiking on animals or on you. They just walk in—they're very thin—and they seem to get through all kinds of spaces."

The good news about the bark scorpion is that it isn't invisible. You might think, actually, I'd rather not go looking for them. But consider the alternative: Would you rather find the scorpion first, or have him surprise you some night in bed?

So what's your strategy for stopping scorpions?

The first line of defense is to cut down on the numbers that come in. Get out the caulk gun. Because the spaces they can crawl through are so small, it may be impossible to block up every avenue of approach. But if you have nothing else to do on a Sunday afternoon, it doesn't hurt to try. Along with caulking cracks, add weather stripping along door seams and screen off any vents. Steps like that will cut down the number of intruders. You'll never develop a completely bug-proof house, but you'll make it less inviting, so there are fewer unwanted guests.

The next step is to evict the scorpions that have already gotten into the house. Where to look? Says Olson, "I've found them all over, but one of the most common places is where there's water, because they drink, too. So when there's a free source of water, like in the kitchen or bathroom, they get in there and they can't get out, because they can't climb that porcelain."

He reminds people not only to look out for scorpions when they're on their Sunday-afternoon purge, but anytime they're in the kitchen or bathroom. "People get into their shower, step on them, and get stung because they didn't look," he points out.

They can also be found on walls. "Most of the time, you have a chance of seeing them, because they cling to the walls; they are climbers," he says.

When Olson finds a bark scorpion in his tub or on a wall, he just captures and releases it. "I just put a jar over them to catch them."

A third line of defense involves shifting their outdoor living areas farther away from your house. You start by going around your house at night with a red-covered light, which scorpions can't see, find the scorpions that are outside around the house, and move them (scoop them up in a jar) elsewhere.

And then you try to push their habitat farther away from your house. Start by removing big rocks; that's a great place for scorpions to hide. (Remember what you were taught in gym: Bend at the knees, not at your back.) Get rid of dead wood and trees and brush close to the house. The idea is to create a big desert—an inhospitable habitat— between your house and their dwelling areas, so the scorpions are less likely to want to venture across that gap.

Dealing with scorpions—as with many other wild creatures—is also a question of attitude, Olson says. "People come to the desert, but they don't want to live in the desert. They won't accept that this is part of desert living." Instead, he says, "You just have to be careful. When you go hiking in the desert, you look where you put your feet so you don't step on a rattlesnake." It's the same with scorpions. "It's a matter of opening your eyes and being aware of your surroundings," not of relying on chemicals to "surround you and protect you from all these nasty, vile things that are out there—that are life."

Olson objects to the use of chemicals as a remedy. In fact, he says some pest control companies play on peoples' fears of scorpions in an effort to get them to use chemical pest-control products, which, in his opinion, "don't do any good." He explains: "You can put down the chemicals, and the scorpions will walk through it, and they'll still come into your house." A Star Wars shield against scorpions doesn't exist. Eventually poisons might kill the scorpions, but not before they've had a chance to explore your house. In other words, they'll still cause you to shriek one evening.

And it doesn't look like these encounters are going to stop anytime soon. True, scorpions aren't at epidemic proportions like some other pests. And, as Olson says, "They're not prolific like aphids or cockroaches." Yet, as humans encroach more and more on the territory of animals, we should expect to encounter our nonhuman neighbors more often. That's especially true with scorpions.

Though scorpions do not bear a lot of young—maybe thirty a year—many of those young survive, because in an urbanized setting, fewer and fewer of the scorpion's natural enemies survive. At the same time, our living areas are providing scorpions with places to hide and with moisture, and are attracting other bugs, like crickets, which they eat. "Everything we do that's nice for us is nice for the scorpions," Olson points out.

"So we make the problem," he says.

## ❖ STOPPING SCORPIONS

1. Heroic weatherproofing.

2. Stop and look first. No bare feet.

3. Clean up around the perimeter of the house.

4. Reduce available water: Fix leaking faucets; don't leave standing water; wipe shower dry.

# CATS, DOGS, AND THE NEIGHBOR'S PETS

There are cat lovers. And there are dog lovers.

But there's a significant minority of the population who are neither cat lovers nor dog lovers. These nonlovers of cats and dogs don't actually dislike these domestic pets; rather they dislike cats and dogs mistaking their yards for bathrooms, lunging at their baby carriage, meowing or barking because David Letterman has said goodnight, excavating their yards (especially when spring's first flowers appear), and cozying up to any leg that happens to be walking down the street. Not to mention giving a free ride to all the ticks and fleas in the neighborhood. Even faithful cat and dog owners aren't too happy with a pet who tortures expensive furniture, or with their neighbor's untrained dog. And what cat hasn't tested the limits of its owner's love by using an antique wicker blanket chest as a scratching post? What dog hasn't threatened to destroy his reputation as "man's best friend" by marking his territory—indoors? Indeed, I've known more than one pet owner who's had to stand guard protecting a new leather couch. (More on defending your furniture later.)

Cats and dogs are animals. Even for people who love them, these animals can be unruly at times.

There's a basic difference between cats and dogs that has nothing to do with size, or voice, or fur. My wife, Peggy, expressed the difference this way:

> To understand how a cat's thinking is different from a dog's, it's useful to understand what psychologists mean when they speak of the difference between guilt and shame. Guilt is an inner knowledge that you've done something wrong. You feel guilty whether anyone else has seen you commit the wrongful act or not. Shame, on the other hand, occurs only when others know that a stain on your honor has occurred. Shame is a loss of face that reflects badly, not only on yourself but on your entire family and your tribe.
>
> A dog can feel guilt, whereas a cat can only feel shame. A dog knows as it jumps onto the forbidden couch that it is doing a bad thing, even if you're not there to tell it so. It may jump up there just the same, but it will still hear the little voice inside its head saying "bad dog, bad dog." A cat will feel no such remorse. As far as it is concerned, if no one saw the crime, it didn't happen. But if it knows you were watching when it misjudged the leap to the endtable and knocked over the lamp, you will certainly see it lower its head and slink off, deeply ashamed, not only for its own clumsiness but for the poor showing it has just made before humans for all catkind.

In quantity, cats and dogs can be a hazard. And cats and dogs (but especially cats) often come in quantity. San Mateo County, California, passed an ordinance that requires individuals—other than licensed breeders—to have their pets neutered starting January 1, 1992. That's how bad the problem of strays is in that part of California (and probably elsewhere—they just did something about it in San Mateo County). The measure was introduced by the county supervisor, who was concerned that the local humane society destroyed ten thousand stray dogs and cats a year. Of course, this law was highly controversial, and as this book goes to press a canine defense fund to oppose the legislation, with $350,000 in the kitty, has been established.

Most local governments are doing well just to battle rabies, which is reaching near-epidemic proportions in many wild animal populations. The majority of local governments rule that both cats and dogs must be vaccinated against rabies, and the Humane Society of the United States recommends that even indoor pets be vaccinated. Tucson, Arizona, recently joined the majority. Starting in 1990 cats had to be vaccinated against rabies and licensed just like dogs. Two instances of rabid cats and thousands of strays gave the city council the rationale to vote for this legislation.

Hamilton County, Indiana, built a $270,000 kennel to take care of its growing stray cat and dog population. There are seven thousand stray animals a year who wander through the county.

Glendale Heights, a suburb of Chicago, also had its problems. Too many cats would roam into people's yards and, according to one resident, "scratch after they do their thing and dig up flowers." Glendale Heights ultimately decided to do nothing about the roaming cats, mostly because it's impossible to regulate cats' behavior. Besides, how do you tell to whom a particular cat belongs? The cat knows, but unless you are willing to follow the cat around all day long to its home (they come home to eat, no matter what wild things they've fed on), you're out of luck. Even if it's wearing an ID tag on its collar, good luck catching the animal and taking a look.

Some places have tried to regulate peripatetic pets, however. Elmhurst, Illinois, enacted a statute that says that not only must cats and dogs be licensed and tagged, but that they are not allowed on anyone else's property without permission (how do you explain that to a cat?). In addition, the law says that no one can have more than three cats or dogs or a combination of both. Trying is one thing—success is another.

Most municipalities have laws prohibiting dogs from wandering onto a neighbor's property or defecating on a neighbor's property. Most cities have leash laws, and some have tried bold—and successful—experiments with pooper-scooper regulations. (Poop-scooping is recycling at its best: In New York City you often watch dog owners reading the newspaper while walking their pets; then just as the dog goes into a crouch, the completed section is quickly placed underneath.) I'm an advocate of pooper-scooper laws; there's something uncivilized—and certainly unhealthy—about dogs defecating in public places. The parasites that dogs deposit with their feces have been the cause of disease in people and in other animals. Neighborhoods where lots of people let their dogs roam freely, pooping where they will, look and smell like medieval villages.

Despite their important contribution toward health, lots of people ignore cat and dog laws. After Senator Robert Dole and his wife, Labor Secretary Elizabeth Dole, revealed that their dog, Leader, was the proud father of eight puppies, the Humane Society protested. (The mother was a schnauzer owned by Senator Strom Thurmond.) When the Doles acquired Leader their adoption contract required that the dog be neutered within seven days. Senator Dole's press spokesman offered this explanation: "[Leader's] not on the loose, which made this a selective breeding. It wasn't like he was roaming the streets of Washington creating litters all over town."

Leash laws are flagrantly ignored, often to the peril of children and other innocent bystanders who are bitten by someone's "friendly" dog. In many suburbs nonroaming laws are also ignored, often to the peril of the dog who's in danger of being hit by a car.

There are lots of cats everywhere. A half hour's drive from Melbourne, Australia, there is a tropical rain forest, the host of many rare species including the fabled lyre bird. The Sherbrooke forest is among Australia's most beautiful places. Unfortunately, the greatest danger facing Sherbrooke forest is cats—cats love to eat what they find in nests, especially lyre bird nests. So the Sherbrooke Shire Council enacted legislation prohibiting cats from taking strolls between 8 P.M. and 6 A.M. Sometimes authorities are even more strict. Ecologists had to track and kill feral cats on Codfish Island, off the coast of New Zealand, because the animals were eating the endangered kakapo, a flightless bird. To be equitable about it, they also killed anything else that might eat the birds, including opossums and stroats.

# CATS

Cats consider their yards and homes as home ranges, and as their owners know, develop favorite special spots where they sit, sun, watch, scratch, hunt, fight, and court. Cats tend to avoid one another, and there is usually a social hierarchy when the animals have to interact regularly in a neighborhood or house.

Cats regard their owners as dominant cat or mother figure, and many of their behaviors toward owners stem from that perception. For instance, when kittens nurse, they also purr to communicate their contentment. Later in life their purr in response to petting, feeding, or scratching may be for the same reason. They may also purr to communicate friendliness. When they nurse, they knead the mother's belly with their front paws. Later in life, the kneading and drooling occur when the cat is getting comfortable in the owner's lap. You can ask your cat to stop and gently press it into a resting position, but they'll always do it if they're inclined.

The most important thing you need to know about cats is that they are very good at being cats. All day *and all night* long they are cats. Here's what I mean. This married couple I know were sleeping one evening. Good activity for the night. Their cat, Sasha, was not sleeping. Sasha was on the prowl and this particular evening Sasha captured her prey: a lovely field mouse. Proud of her accomplishment, Sasha brought the mouse, now dead, to her owners' bedroom. They will be so pleased, Sasha must have thought (in the way that cats think). Sasha put the mouse on the pillow next to her female owner's head. The woman screamed. She woke her husband who, half asleep, picked up the mouse, opened the bedroom window, and tossed the mouse back where it came from. Unfortunately, they forgot that their windows were alarmed—a mistake anybody could make at 2 A.M. They went back to sleep.

Normally, this kind of cat and mouse story would end here, but this isn't a normal story. There was a minute delay on the alarm and the couple were both hearing impaired; they slept without their hearing aids. The alarm went off—silently to them—but not so silently to the police. The couple only awoke when the police headlights were shining through their window. Imagine: They thought their house was being robbed. If I were them, next time my cat deposited a mouse on my pillow—I'd still throw it out the window.

Males scent mark by spraying an object with an aromatic mixture of urine and fluid from anal scent glands. Natural sexual cycles, independent of the female cycle, lead them to scent mark more at certain times than others, but a female in heat certainly stimulates the behavior. Unlike dogs, they will not spray over another male's marks.

When they rub their faces and tail tips against our legs, they are scent marking us with secretions from glands on their faces and tail tips. (And you thought it was adoration.) They may scent mark another cat using a gland located on their foreheads.

Males will fight over territory at certain times. Like most animals, their ritualistic behavior is an attempt to ward off actual fighting. They may adopt the posture of Halloween cats, with arched backs, raised fur, and erect ears. As the tension escalates, they may drop to the ground and roll over on their backs to better strike with all four paws. A cat fight ensues with great noise, but the posturing often prevents a fight.

We perceive cats as being playful, but much of their play is a displaced prey-catching behavior. They may leap at nothing in the air, attack bits of paper on the floor, or dash about for no apparent reason.

When they catch prey, they will put the animal down as soon as it stops moving. Then the cat will groom itself, pick up prey, walk a short distance, and put the prey down again. They like to look good for dinner. Eventually the cat reaches a private place, and if the prey hasn't escaped, the cat will eat it. Simulating a mother cat's behavior of bringing home food to kittens, a cat may bring its bounty home to its owner. And most pet cats should be pretty unfamiliar with kittens because their owners have had them neutered or spayed.

Cats usually breed once or twice a year. Each breeding season, the female will be in heat several times if she has not already mated. Only the sex act stimulates ovulation. The male will breed at any time. Male cats use their sensitive noses to find females in heat. They have a short courtship, and they may mate several times in succession, but they don't form a pair bond. The female raises her kittens alone.

Gestation lasts sixty-three to sixty-five days. A few days before the event, the mother seeks a dark, secluded spot to give birth. For the first few weeks, kittens sleep and feed, and the mother occasionally leaves the litter to find food for herself. At about seventeen days, kittens can walk a little and at five weeks they begin weaning and eating solid food. At eight weeks they are on solid food and no longer depend on mother's milk. They then begin to learn to hunt and have it mastered in two or three months.

Even if you love *your* cat, you may not be crazy about somebody else's cat. So says Marta Vogel of Takoma Park, Maryland. One evening she heard some loud noises in her basement. "I was used to sounds in the basement because I have two cats myself," Marta said. But this was a different noise. "This sounded like a drunk tearing through the basement. I called the police." The dispatcher told her to stay on the line. Finally, the dispatcher said, "Okay, the police are at your door." Marta let two police officers in; they proceeded with guns drawn into her basement. "The first cop went around the corner and ran into the neighbor's cat who was frantically trying to figure a way out. That cat had gotten in through my cat door."

What can you do about a neighbor's cat wandering in your house through your cat door? Almost nothing, short of changing the lock. No kidding. You can purchase a magnetic cat door that will open only in response to a special collar worn by your cat. (See the resources section for suppliers.)

But what about the neighbor's cat wandering through your garden or on your porch—or worse, around your bird feeder? For this, there probably is nothing you can do. In many localities, cats are considered feral creatures and are allowed to wander where they will (but check your local laws). If a neighbor's cat truly is a problem, consider erecting a fence around whatever you want to protect. Of course they can climb it, but the fence may be enough to encourage the animal to hang out somewhere else. With a nice enough neighbor, you might get the neighbor to pay for part of the fence—or to keep the cat in. It is, after all, in the owner's interest to keep the animal in.

One reason being that there are cars on the streets, driven by people who won't even stop after they run down your pet. Or even worse, there are cars driven by people who are looking out for animals; they're the ones looking for strays to sell to an animal lab. Cats can also make other cats sick: Feline leukemia is transmitted cat-to-cat.

If you can't persuade your neighbor to keep the animal in, you may keep a cat off your property with an animal repellent spray. Get Off My Garden is designed to keep pets and wild animals away from your plants and patio because the smell is so awful. We don't rely on our sense of smell as much as cats do, so we won't notice the smell.

If you're worried about birds at the feeder, baffles designed to keep squirrels away will also keep cats off the feeder. Place your bird feeders and baths in open areas, away from cover that could hide a crouching cat. Finally install a "bird guardian" in the openings of your bird-houses. The devices prevent animals from reaching into birdhouses and snagging helpless birdlings. Bells on the cat's collar won't warn birds in danger because the cat will learn to walk so the bells don't jingle.

A note on Lyme disease: Check your outdoor-indoor cat regularly for ticks and remove them by pulling them out with a pair of tweezers. Signs of the disease in your cat include lameness and a recurring or persistent fever. Have the animal's blood tested to see if there is a high antibody level for Lyme disease. Flea and tick collars pumped full of toxins expose your pet to unnecessary danger, so choose alternatives. See the chapter on bugs for more information on ticks.

And finally, what about furniture? Not being a cat owner (though I am friends with plenty of cat owners), I never really appreciated the potential seriousness of mixing cats and new furniture, until my mother-in-law told me that the fabric for her new couch was decided by her cat, Emma. I was perplexed. Emma? Yes, Florence replied, Emma likes to scratch certain fabrics but not others, so that's how we're going to choose.

Cats must scratch something, and they will scratch your new couch if you don't provide an alternative. And there are many alternatives. Scrap carpet turned so the rough bottom faces upward; expensive, fun, carpet-covered jungle gyms from the pet store; sisal-covered boards hanging from a doorknob; a plain log on the floor. Just be careful it doesn't approximate something already in the house, because it's hard for some cats to understand the difference between the shag carpet in the living room and the shag carpet all over the jungle gym.

A friend of mine sent the wrong signals to her cats and ended up with a destroyed arm chair. When she noticed her two cats were picking on the arms of the upholstered chair, she pinned plastic over the arms, anticipating the cats would find it unappealing in texture. She returned after a workday to discover the stuffing on the floor. Only then did she see the connection her cats realized immediately. The plastic on the armrests sent the message: "Scratch me," because the cats' litter boxes were lined with plastic, and the cats routinely shredded the material when they visited the box. She invested in a sisal pole on a stand. No more problems. There are also sprays designed to keep pets away from furniture, but you may find the smell offensive as well and end up sitting on the floor with Fluffy.

Houseplant-eating cats are annoying and in danger. Foul-tasting sprays will stop the habit—try a homemade concoction of hot sauce and water or a store-bought mix. Digging in large plant pots, and worse yet, using them for a litter box, is a tough habit to break, because you'll rarely catch them doing it. If they've used the plant as a bathroom, you'll need to repot it and throw away the old soil. Then try keeping them away from the plant. Stretch plastic wrap across the top of the container, covering all the soil and leaving only enough space for the plant stem to poke out. Secure the wrap with tape. After about a month, your cat should break the habit. Be sure to watch the plant soil for signs of mold during treatment. Wire screening or mesh is also a good excluder.

Cats like to nestle in cribs. They like to play with the baby's toys and they like to nuzzle up against your newborn. One of the best ways to keep a cat out of the crib is to put crib netting above the crib. This netting is sold in baby product stores and catalogs.

Sometimes your otherwise good cat can become a problem-cat. You may know the type: hisses arbitrarily, pees on furniture, fights all the time with other cats—generally unpleasant. Believe it or not, Valium may be the solution. Your veterinarian can tell you whether Valium will help. I've seen cats on Valium. My brother-in-law had a cat, Laska, that frequently urinated on furniture. Cat-sized doses of Valium solved this problem. It also made the cat move in slow motion—very amusing. (The cat seemed to enjoy this, too.)

If your cat is having problems remembering to use the litter box, maybe you should consider stepping up your litter box sanitation duties. Scoop solid waste daily and change the litter entirely two or three times a week. (Similarly, recurrent urinary tract infections often result from not-so-tidy litter boxes.) Or you might toilet train your cat. Paul Kunkel trained his, and then went on to write a book about it, *How to Toilet-Train Your Cat: 21 Days to a Litter-Free Home*. But it won't be easy. You gradually raise the height of the litter box and accustom the cat to the toilet seat. While Kunkel seems to think you can be litter-free, other trainers have found that you can only train a cat to urinate on a toilet, and that you'll have to keep a litter box around.

Despite their reputations for surly attitudes, cats can learn and obey all house rules. Just like your friends and relatives, cats won't be instructed with loud demands and brute force. And unlike your friends, they don't make the connection between the slashed drape and your angry words several hours after the incident.

Some people use negative reinforcement. Several mail-order catalogs carry an electrically charged pad that lightly shocks animals who step on it. It can keep cats out of a room or off a piece of furniture, and there are even outdoor versions to keep cats off the car. You can even order dummy mats so the reinforcement is still in place when you use the charged mat elsewhere. These mats may be successful in keeping the cat out of your baby's nursery.

Or try a low-tech solution of a squirt bottle (or water pistol) and a firm, low voice issuing simple commands: "No." This one takes some time, plus you always have to have the bottle with you to be ready to discipline. It has an unfortunate side effect. Eventually the cats wise up and see the bottle as the enemy. Then they run when you do the ironing or get out the window cleaner.

Cats respond to your voice. They learn what "No" means just like they learn "Good kitty." Say it like you mean it, and use different tones for different messages. Sure you'll repeat yourself a lot; they don't want to hear discipline, and they can't get enough praise. A mother cat would discipline a kitten with a hiss and a firm but gentle push against the forehead of the cat. Since they regard us as big, furless cats, this works too. In time, they learn walking on the kitchen counter, sleeping on certain chairs, climbing drapes, and scratching screens all elicit disapproval, "No," and hisses from you. And they stop, needing only occasional reminders.

The key to disciplining your cat is consistency and reward. Look for training books that emphasize positive reinforcement, not physical punishment. I've listed several of my favorite cat books in the resources section.

More than half of all United States households have pets, so the variety of available pet products will stimulate your imagination. Mail-order catalogs carry everything from special cat hair pick-up rollers to cat litter that will alert you if your cat has a urinary tract infection. See the resources section for leads.

## ❖ COPING WITH CATS

1. Try repellent sprays and fences to keep them out of your yard.
2. Get a big dog.
3. Think like a cat.
4. Get a good scratching post.
5. Remember that neutered males rarely spray.

# DOGS

All dogs belong to the same species. They are social animals and treat humans as pack leaders or parents.

We recognize friendly behavior in dogs: backward pointed ears, low tail wagging, happy barks, and lifting of one forepaw. When we scold them, they put their tails between their legs, lay their ears flat, and maybe lie on one side, exposing their genitals in a passive submission display. We also recognize dog aggression: stiff legs, raised hackles, alert ears, and low growl with lips pulled back. We probably miss scores of other messages because we're human.

Dogs are good guards because they're territorial. Males especially make good guards because they will fight in disputes over territory (females are less apt to fight, except to protect their puppies). Males tend to scent mark their house or yard.

All dogs will scent mark an object, even over another dog's marking. Scent marking helps make the area familiar to the dog and assists in communication, telling other dogs who has been there. Dogs have keenly developed senses of smell because they have an organ of smell in the roof of their mouths in addition to their noses. The olfactory area of the dog's brain is larger than that in a human's. When dogs dig up the ground where they have just urinated or defecated, they release scents into the air in addition to giving a visual sign to the visit. Although we don't understand why, dogs may roll in foul-smelling messes, and that in itself is a good reason to keep them confined in the yard. Dogs have a good sense of hearing, and they hear higher frequencies than we do, which is why they often react to a sound before we hear it, but dogs are color blind and don't see as well as humans.

When it comes to making more dogs, the most important sense is that of smell. Males find females in heat by picking up their scent in the air. Female dogs usually come into estrous two times a year, and males will mate at any time of the year.

Sometimes a pair of dogs will become stuck together during mating. It's cruel to abuse them when this happens and to douse them with water or angry words. Dogs will remain connected like this for ten to thirty minutes, which probably helps ensure fertilization. However, you may want to discourage other sexual behaviors through negative reinforcements, such as yelling or using a squirt gun. It's not uncommon for dogs to engage in sexual behavior with inanimate objects, familiar cats, or their owner's legs.

Puppies are born after a fifty-nine-to-sixty-three-day gestation, and a mother may give birth to anywhere from one to twenty-three puppies, depending on the breed. For the first two weeks, the puppies do little else besides eat and sleep, and their eyes open after three weeks, although they can't see well at that time. During the third week, puppies begin to play with their littermates, and they are weaned at eight to ten weeks old. Between their fourth and twelfth weeks, they begin to develop their socialization skills and engage in mock and real fights with their siblings. Life span and onset of sexual maturity differ from dog breed to breed.

Dogs are social and curious animals, but you don't have to be similarly social. Even the tamest dog can be dangerous. Untame dogs, almost by definition, are always dangerous. You should consider any dog with which you are not familiar to be untame, and any dog that is unfamiliar with *you* is dangerous. Every year, approximately forty thousand Americans suffer bites from dogs. You should keep your distance, and you should especially keep your children at distance.

As the human population increases, people who let their dogs roam are doing their pets a disservice. They get into trouble, anger the neighbors, and sometimes get hurt.

In November 1990 two huskies that had been let off their leashes broke into a pen holding several deer at the Ecomuseum in Montreal, Canada. The dogs crashed down the gate. Three deer escaped and one fawn had to be put to death because its leg was badly broken. The dog's owner said that he usually let his dogs run free in the woods near the museum. After this incident, the dog's owner said he would make amends: "I won't let the dogs run free for a while." How thoughtful, museum officials thought. One official remarked, "He wasn't apologetic at all." According to the museum, the owner had been told about the dogs wandering into the museum twice before; each time the dog's owner did nothing. And you wonder why you can't get your neighbor to keep its dog from pooping on your front yard.

Many towns have ordinances prohibiting dogs at public events—mostly because they are unpredictable and potentially dangerous. Roselle, Illinois, is one such town. Roselle's law was passed when the town's trustee saw a resident walking two pit bulls through a food festival on Main Street. Much as the Red Sea parted, people moved away from those dogs as the owner strolled along.

Most dog owners, like most car drivers, believe that their rights are absolute: Dogs should be free to roam wherever they want and do whatever they want. (Cigarette smokers and hunters have the same point of view about their activities; unfortunately, this "It's my right" attitude is invoked by most of us to justify some behavior of our own at one time or another.)

Roaming animals aren't just a problem for rural areas. In Manhattan a German shepherd was killed by a subway train. It seems the animal somehow found its way into the tunnels and eluded officials for hours before eventually being killed.

If you can't get your neighbor to control his dog, call the local authorities. Most localities have leash laws, and animal control officers will ticket the owners. The general principle is: A dog must be under its owner's control at all times. But before you declare war on the intruding dog, at least talk to the owner. When you decide to throw rocks at a stray dog, or spray it with a hose, or yell at it, you may discover too late that you're abusing a fifty-pound sociopath with sharp teeth. If you're concerned about encountering stray dogs away from your home, invest in some mace for animals or even one of those three-foot sticks that will deliver a nasty shock. Ultrasound devices designed to repel dogs are highly effective in getting rid of an occasional stray, says the Humane Society. The loud, high pitch is painful to the dog, and it will flee. Still, it won't work twice a day on your neighbor's dog.

If you have small children, you shouldn't tolerate any loose dog: Wandering dogs pose a grave danger to children. This is particularly true with large dogs or aggressive breeds. In addition, "some dogs are genuine sociopaths," says one dog trainer. These dogs will attack for no apparent reason. Talk with your neighbor right away about the problem. If the neighbor won't relent—and many will not—then take whatever action you must to protect your children. By all means, call animal control. Call the police. Post signs with that dog's picture in the neighborhood warning others about danger. If you see the animal wandering, tie it up and call the pound: That usually gets the message across.

Dog walkers can be another nuisance. For some reason, some dog owners regard your front lawn as a toilet for their dog. You've seen this at one time, I'm sure: Someone walking their dog while reading the newspaper, just letting the dog go in its favorite yard. In my opinion, it takes considerable nerve to do that. If somebody's dog has selected your yard as its favorite, then there are several steps you can take. First, talk with the owner. Hang out on the front porch—or stand stealthily at the front window—and confront the person. Most of the time, that works. If talking doesn't work, yell next time. If that doesn't work, take a photograph of the event and tell the owner that next time you see his dog on your lawn you'll deliver the photo to the police.

Even if you can't control your neighbor's dog's toilet habits, you can control your dog's. It's fine to keep your dog confined to the backyard, but clean up the solid waste regularly. You can even train the dog to use just one part of the yard. Leash the dog before you let it into the yard and then walk it to the designated area. When the dog does its business, lavish praise upon the pet, maybe even give a treat. Follow this regimen for several weeks. Then let it out with a leash and without you. If you see the dog using an undesignated area, guide it by leash to the designated area and praise it, even if it doesn't use the bathroom. Your dog will soon learn where the doggie bathroom is. Clean up the solid waste daily. You can even get a small septic system for dog droppings, the Dooley.

It's a good idea to have a fenced area for your dog, something that keeps it in and keeps out other visitors. There are chain link fences, picket fences, and even invisible fences. An invisible fence, one that is installed underground and is linked by a radio frequency to your dog's collar, shocks the animal when it tries to leave the area. Well, they're easy on the eye, but consider this: a person, a bigger dog, a wild animal, even a rabid animal can cross the fence at will, while your dog will suffer a shock.

If you can't keep stray dogs out of your garbage by placing garbage cans in a stand so they won't tip over, try spraying a distasteful liquid on the bags: Alum, hot sauce, or cayenne pepper tastes pretty bad to dogs, or try a commercial concoction. Stop dogs from chewing on your shrubs and young trees with commercial sprays like Get Off My Garden or a homemade spray mentioned above. You can find similar sprays for indoor use. A bitter apple spray is nontoxic and distasteful to dogs.

Dogs who like to chew need an alternative, though, so be sure to provide sturdy toys while you're gone—and remove temptations like shoes, handbags, and socks. A good, if messy, treat is a bone filled with cheese spread or peanut butter. Dogs get bored and lonely, too.

The retail industry has plenty of solutions for barking and jumping, although barking is a part of the dog's nature and only so much can be done to curtail barking. Because dogs are sensitive to high-frequency sounds, many disciplining tools rely on them. You can get ultrasound devices to repel strange dogs, contraptions that sense dog barking and emit ultrasound to silence the dog, and hand-held ultrasound machines to use in training your dog. Unfortunately, nothing is that easy.

"I'm not impressed with them as a device that can be used alone," says Peggy Pachal, canine behavioral consultant with the American Dog Owners Association. "It's not a negative stimulus unless combined with other methods." She pointed out that used alone, the sound is just a noise that can be blocked out, like screaming mothers or traffic. Different dogs may react in different ways: Some dogs may freeze and look, some may ignore the sound, and some may even react aggressively. Distraction is not a correction device, and the machine can't do all your work. But Pachal said that ultrasound devices are used by professional trainers in combination with other behavior modifiers, and that in time, a dog could learn to respond to the ultrasound device. In the end, they're devices to use in your training program, not in place of it.

The key to disciplining your dog is consistency and reward. Look for training books that emphasize positive reinforcement, not physical punishment. I've listed several of my favorite dog-training books in the resources section.

More than half of all U.S. households have pets, so the variety of available pet products will stimulate your imagination. Mail-order catalogs entirely serve your every pet care need. They have everything from special doggie life-jackets to heated sleeping pads for older dogs. See the resources section for leads.

## ❖ DIVERTING DOGS

1. Use repellent sprays to keep them off your property.
2. Fence your yard to prevent your dog from wandering.
3. Provide plenty of chew toys.
4. Talk to the owners of a misbehaving dog before you take action.
5. Carry Mace and use it if necessary.

## ROOSTERS

Most people who have pets have cats or dogs. Oh, a few have parrots, and probably plenty have tropical fish—but these are indoor pets that you never see.

But from time to time, it strikes a family's fancy to purchase an exotic pet. Not necessarily exotic by virtue of the animal coming from a distant country, but exotic in terms of being unusual. For example, my friend Caroline Petti had chickens when she was a child. Actually, the chickens started out as adorable little chicks, but of course that didn't last too long. Caroline's chickens, even as they matured, didn't bother anybody, so she was able to keep them for a long time.

Not so with other exotic pets: Some can be exceedingly annoying to have as neighbors. Take roosters for example. Portland, Oregon, which allows residents to own rabbits, pigeons, chickens, bees, sheep, and even goats has an ordinance (Section 13.16.075 of the city code) against roosters. Why were roosters singled out? The answer is that roosters are awfully nice, but only if you enjoy seeing the sun rise.

The first step to take when your neighbor has an objectionable pet is to check the local ordinances. If you're lucky, whatever they have—a brown bear, a rooster, a laughing pig, a cat that seems suspiciously larger than your run-of-the-mill tabby—will be illegal. For the most part, cities act swiftly to expel illegal animals.

But sometimes the law isn't all that clear. Or rather, the law may be clear but the species you're concerned about may not be so obvious.

Let me explain by way of example. An Oregon family moved from the country back to the city, Portland, in fact. They brought with them a potpourri of animals, but, cognizant of the law, left their roosters behind. Because chickens are permitted in the city, this family brought with them four hens—fancy, pedigree hens. One was a Polish Crested hen, another a South American Rompless, the third a Black Australorp, and the forth a Silver Spangled Hamburg. Fancy chickens, but definitely chickens. Not long after moving to Portland, the head of the household awoke at 5:10 to what he described as a "kind of pathetic cock-a-doodle crowing." He looked around, but couldn't find a rooster. The next morning, the same thing—and so on for days and days. Finally, the family discovered that the Black Australorp, a female, had decided she enjoyed impersonating a rooster. The "pathetic crowing" soon developed into a robust cock-a-doodle-do. Meanwhile, the law doesn't say anything about rooster-impersonating hens, so the neighbors are out of luck. But they have this consolation: The family that owns the Black Australorp suffers, too.

# COYOTE AND
# MOUNTAIN LIONS

People living in what I call the fringe suburbs, those localities that are on the rim of urban areas butting up against open areas, see a wider variety of wildlife than urban residents—including large carnivores like coyotes and mountain lions. As the suburbs sprawl and encroach on the wildlife's range, these animals are going to be more interested in nibbling on foods they ought not to eat, or on things that are not food at all.

## COYOTE

In the summer of 1991, members of Spokane, Washington's suburban North Side were demanding that coyotes in the area be trapped and killed. Cats were disappearing, and one resident found her cat's head and shoulders about seventy-five feet from her house. Officials declined to do anything; they couldn't use traps, shotguns, or poisons in such a highly-populated area, so they waited for the natural high mortality rate of coyote to solve the problem, and warned pet owners to keep the animals confined at night, when coyotes do most of their hunting. Residents were outraged.

Defenders of Wildlife's senior wildlife biologist, Al Manville, looks at the problem from a slightly different perspective. "Which is the native species and which is the introduced one? The coyote was there first. Responsible pet owners take steps to prevent coyotes from getting to their pets."

It's a reasonable enough solution. Leaving your pets unrestrained in coyote country is like filling up a bird feeder; animals will come to eat.

Through 1972, when the federal government banned it, Compound 1080, a poison, was used by ranchers to keep coyotes at bay. Since 1972 ranchers have had to rely on mechanical traps, cyanide, and guns, methods they insist are not as effective. In addition, the decimation of the wolf, the main competitor of the coyote, has opened up habitat for the coyote. The coyote population, especially in Texas, has rebounded; you can now spot coyote around the Dallas–Fort Worth Airport. Houston has its own set of problems: Coyote have been eating the geese and ducks that live on golf courses. They've made their way East as well, though they were originally western animals.

Coyote are making a comeback in semirural areas. In 1990, trappers caught over a dozen outside of New York City. Through the late 1800s coyotes were principally residents of the American West. As we killed wolves, the coyote took advantage of this ecological gap and began to move east, appearing in New York state sometime in the 1940s.

Not too far from New York City, coyote are developing a stronghold. The Westchester County Airport considered trapping a family of coyotes living near the airstrip, but then decided against it, feeling they were really harmless. Coyote numbers are up in the county, and why not? Prey animals thrive around humans, so why wouldn't the coyote come along to take advantage of the good eats? There are rodents, rabbits, carrion, birds, bugs, an occasional deer, and pets.

In many parts of North America you can talk with coyote. A loud *Yip, yip yaheeee* through cupped hands may encourage a response from the real thing. Calling coyote works best in the spring when the animals are forming mating pairs.

Not all people dislike coyote. Their doglike manners, mellow howl, fluffy tail, and bright eyes make them interesting, almost endearing animals. Some gardeners want coyote around, because coyote keep fruit- and vegetable-eating mice and rabbits in check. Plus they keep deer at bay. Coyote are not completely trustworthy in the garden though; some of them seem to have developed a taste for melons and will occasionally help themselves.

Fortunately, coyote are relatively shy and not interested in people. When on the prowl, it's hard to notice that they're even nearby. If you were looking you could see them anywhere, in most of the United States, Canada, and Mexico. They're usually pretty small animals, with males weighing about twenty-five to thirty-five pounds and females being somewhat smaller, though they can be much larger or smaller. They're mostly gray in color with a white belly, although sometimes they vary in color from red to black. Wildlife experts suspect that some coyote are actually coyote-dog or coyote-wolf hybrids, especially in the East, which may account for the wide variation in size and color.

Coyote mainly enjoy rabbits and rodents, but their diet includes vegetation, carrion, and garbage (if that's all that's around), and livestock or small pets (if that's all that's around). In the coyote's defense, a garbage mess or what appears to be a coyote attack may often be caused by feral dogs. Coyote are adaptive feeders, one reason they thrive in many areas. When they live around humans, they hunt mostly at night and during the early morning hours. In more isolated locations, they may be active around the clock.

Coyote are adaptable to changes in the environment too and can thrive in forests, deserts, mountains, grasslands, and the bush. Unlike many animals who perish when their sources of food disappear, coyote move with their food.

Coyote are found in almost every state. And they're resilient, tough, and fast—some can run up to forty miles an hour. Most of the year, coyote life centers around the dens if they're raising young, but at other times, they hide in dense brush or other sheltered areas and come out to hunt.

A coyote litter consists of six pups; the breeding season is late winter or early spring. Males and females raise the young together. Sometimes two females will share a den, having mated with a single male. The dens are secluded, maybe in rocky areas or in a den abandoned by a skunk, badger, or other animal.

Before the pups are weaned, at six weeks, they begin to eat meals regurgitated by the parents. Within two months, they are ready to go on short hunting trips. By the fall, the family usually splits apart, although they may stay together until the next spring. Females are often ready to breed before their first birthday. Some biologists believe that eastern coyote exhibit more wolflike behavior by living and hunting in packs rather than living the solitary lives of western coyote. John Anderson, a trapper in New York, says, "They look like coyotes and respond like wolves."

When coyote take livestock, people get pretty worked up and demand action. Anderson finds that coyote are a lot like people—it only takes one troublemaker. When he must take action to prevent damage, he watches the pack to find the one that attacks pets or livestock, and kills it. "Make sure you don't get the dominant female," he says. The others disappear until the next season. "You don't need to take out five or six. You take out the troublemaker." Anderson is licensed by the state to do trapping and hunting of problem animals; so don't try this at home.

Keeping the coyote at bay is a little tough. At first scare tactics like loud radios, propane cannons, and flashing lights may keep the animals away. But then they get used to it. They're sneaky. If you notice an individual coyote constantly watching a house or casing the neighborhood, you may have a problem animal. But Anderson says, "You can't just go out and hunt them, it won't work. They're not all bad, and most are pretty good. It's the ones who start to watch the house waiting for a meal that make the problems."

Locations along washes, abutting open land, or creeks are prime coyote habitats. Larry Manger, a wildlife biologist with the U.S. Department of Agriculture in California, says, "In most cases, a coyote needs a cover to work a neighborhood, but I've seen them walking down the middle of the street." They've been around long enough to become bold. If nobody is picking on them, they just continue to expand and push.

While no one seems to advocate killing all coyote, instead they suggest behavior modification—on the part of humans. Manger says, "If you live in an area with coyote, I recommend if you have smaller dogs, or especially cats, keep them in. My feeling about domestic animals is they shouldn't be running [loose], period."

If you have an enclosed yard, be sure there aren't places where the coyote can get under the fence. As a rule they won't go over, they'll go under. Lots of times they'll just dig themselves out a little hole, about the size of a basketball. Manger says, "Just so they can get their head under and pull themselves through; they don't need much." So always monitor your fence line to be sure it's secure. But don't be surprised if an enterprising individual climbs your fence. (Electrifying the fence will offer added protection.)

Or if you let your dogs and cats out in questionable areas, stay out there with them and supervise. If your dog is large and aggressive enough, it may serve to keep coyote away from your home.

But don't let that big dog roam. John Anderson reports that a male doberman in Tenafly, New Jersey, escaped from its owner for several days one spring, bred with a female coyote, and brought her home. The female waits for the male when he goes into the house to eat, and when the man walks his doberman, the female and the pups follow. The doberman father calls to the coyote with high-pitched squeaks and the female and pups come along. Anderson says, "She won't attack the man, because the doberman accepts him." Now that the dog is being shipped away to California, the coyote will no doubt disperse.

Just because you stand at the door while the dog is in the yard doesn't mean the coyote won't attack. Coyote work very quickly. Manger says, "I've had people let their dogs out in the morning—two minutes and they're gone." They don't care if the pet is a thousand-dollar show dog or a much-loved cat that's been in the family for eighteen years.

Coyote will commonly travel a long way for a meal. Manger details a coyote-killed pet case he investigated. "When the woman called me, I was a little skeptical, because I took a look at the map, and I thought, 'My God, that's a long ways from nowhere. It's all residential.' So I went by, and well, her back fence was down. There was an easement back there where the phone lines ran, and I followed it for almost a mile. It ended up opening out into a large field, which then opened out into the river. So this coyote worked his way along this easement, probably looking for dogs and cats, and probably had been doing it for quite a while." Usually when there is a coyote in the neighborhood, Anderson starts getting calls from people who are missing dogs and cats and want to know if he's trapping around the area. (If he catches any domestic animals, he either frees them on the property or releases them to the pound so their owners can come get them.) But missing pets usually indicate that a coyote is around.

Coyote make great plans. They'll watch a household to get the schedule down and then one morning when the owner lets the dog out at the same time as on the previous five days, the coyote swoop down on the pet and kill it before the owner's eyes. Anderson says, "They make a fox look like it's retarded." Some hungry coyote hunt during daylight, so it's never a good idea to leave your small dog confined to the backyard while you're away. Or, if you do leave your dog out, give him a key to the house, that is, a special doggie door that opens only when activated by a radio frequency on the dog's collar.

Coyote are occasionally dangerous to humans. Naturally, it's the height of stupidity to put a piece of food in your hand, approach a coyote and say "Here, doggy, doggy." (Actually, I take that statement back: Performing this trick with a grizzly bear is the height of stupidity; but doing it with a coyote is only one notch down the stupidity ladder.) Coyote can carry rabies. Almost worse, they can bite a big piece of meat off you. Coyote are aggressive animals, but generally leave humans alone, unless you try to feed them. Truly wild coyote will leave people alone; the habituated ones are most dangerous. A coyote who's gotten a handout once will learn to expect a handout twice.

When coyote start taking pets, people naturally think to protect their children. Children are killed by wild animals, but rarely. It often seems that someone always knows about a mauling or killing done by a coyote in another state or to a great aunt's best friend's second cousin's daughter. Unattended infants and toddlers could be targets for a bad-egg coyote (or a bad-egg human), so never leave children alone outdoors. Older children who are not large enough to discourage a coyote (that is, who are under forty or fifty pounds) should play close to home and come in at dusk.

Still, keeping coyote away can't hurt. If coyote are bothering your trash, keep it tightly sealed (a good idea in any environment, because many mammals are trash hunters). The key is to keep the cans upright; you want to make sure that there's no way a can will overturn— the coyote's objective. Changing the aroma of the garbage can help, too, by adding moth balls, pepper, or ammonia to the can.

Coyote also are attracted to your dog's food. If you live in coyote country, don't feed your pets outside. Even if you bring the bowl in, there's still residue. Dogs are sloppy, they leave the scent on the porch, leave a few crumbs. Anderson says, "When things are bad, coyote have been found with ten pounds of cow manure in their stomachs in the middle of the winter. So they'll eat anything. Even if it's just a few licks off a piece of wood."

Some enterprising individual could probably live-trap coyote, but relocation would be a problem, since coyote seem to be living in most available habitats. And more would just move into the cleared one.

## ❖ CAJOLING COYOTE

1. Don't let animals (or children) travel alone at night; confine animals in barns, sheds, or enclosed pens overnight.

2. If you have to let pets out in the evening, walk them.

3. Keep the brush around your house cut down close; keep the yard clean so it doesn't provide cover for rabbits, mice, or squirrels, giving the coyote reason to hunt.

4. Keep your garbage secure so it's not an advertisement to feed at your house.

5. Don't feed coyote, that is, don't feed your pets outside.

## MOUNTAIN LIONS

In the western states, mountain lions (aka cougars) come into increasing contact with humans both because people are moving to isolated areas and because their towns are encroaching on prime cougar habitat. As with coyote, most complaints center around missing pets and attacks to livestock.

Mountain lions are considerably larger than the coyote, weighing between eighty and two hundred pounds. They are the largest cat native to North America, and range from western Canada south through the western United States and into Mexico. They are also found in the

Southeast. They range in color from gray through blondish to cinnamon, and the young are yellowish with irregular spots.

People rarely see mountain lions because they're shy and mostly nocturnal. Solitary and territorial, their populations aren't very dense. The average home range is ten to twenty square miles, although they'll travel much farther than that to hunt if they have to. They often surprise their prey by attacking from above, from a tree or rock outcropping.

Mountain lions are usually born in the late winter and early spring, although they may breed at any time. A female bears one to five young and raises them by herself. The kittens nurse for about three months, though from the age of six weeks, the mother will supplement her milk with meat. Litters are born about two years apart. Mountain lions will live about ten or twelve years in the wild.

Mountain lions eat differently according to where they live, but prey on just about any animal they can kill: everything from mice to moose, including elk, bear cubs, mountain sheep, birds, rabbits, coyote, fish, and the occasional pet. They rarely prey on livestock, and statistics from western states show that mountain lions account for less than twenty percent of the losses of livestock to wild animals.

Hunted by bounty hunters and driven out of their habitat by human populations, the mountain lion population was in danger not too long ago. They're beginning to adjust to us—and us to them—and now the population seems to be stable, but not thriving and expanding like that of the coyote. Many states still have a mountain lion hunting season.

## ✤ THWARTING MOUNTAIN LIONS

1. Remove brush and tree cover that the mountain lions may use to get close to your house and outbuildings. Clear up to a quarter-mile from the area.

2. Confine pets and livestock at night.

3. Bright lights, loud music, or dogs may repel mountain lions, but not forever.

# DEER

Deer are among the best known of our large mammals. According to wildlife experts, the most common types of deer include the white-tail (found throughout the United States with the exception of some western states) and the mule deer (found primarily in the West). The white-tail is so successful in the East that it's slowly migrating into western regions where it hasn't lived before. In addition, there are several subspecies, like the tiny key deer (found in the Florida Keys), the smallest of North America's deer. (Talk about cute!)

Based on my own encounters with deer, however, I have developed a somewhat different classification system. But more on that later.

With the exception of the male key deer, which weighs only about fifty pounds, adult male deer weigh between two to three hundred pounds. Does weigh from twenty-five to forty percent less than bucks.

Deer tend to live at the edge of forests, browsing in open areas. They use the denser forest areas as cover—for shelter in winter and for escape from enemies. Deer especially thrive in agricultural areas, where fields are interspersed with woodlots and streams. In addition to the natural food, they thrive on our crops: corn, alfalfa, soybeans, vegetables, fruit, and grain.

Deer are herbivores, feeding heavily on the bounty of summer and fall and making do in the winter. In the summer and fall they eat flowers, shrubs, vines, persimmons, acorns, and cultivated plants like corn, apples, and ornamentals. Occasionally they eat grass. In winter, deer eat mostly twigs, bark, and the occasional evergreen. They feed most actively at dawn and dusk, when there is less danger; at midday, they are likely to be bedded down in a secluded spot, chewing their cuds. At night they sleep in secure tall grass or dense brush.

Deer change their coats to match the season. At birth, fawns are rust-colored with white spots, which offer camouflage. Adults sport a reddish-brown coat in summer, which changes to a gray-brown fall and winter coat.

Another thing that changes with the season, in males, are antlers. Bucks grow antlers from April to August. The antlers are nourished by a layer of soft, vessel-filled tissue on the surface called "velvet." Different deer have differently shaped antlers: those of the mule deer are forked while the white-tail's tines rise out of a central beam. Both mule deer and white-tails shed their antlers in midwinter.

Males live in small groups of two to five most of the year, females, or does, in larger groups of two to nine, including offspring.

Depending on which part of the country they live in, deer breed between October and January. One buck may mate with a number of does; the animals do not pair off. After a gestation period of about six-and-a-half months, does give birth to fawns, usually in May or June. While deer typically bear twins, the number of fawns relates closely to how abundant the food sources have been. Under optimum conditions, females may mate at a year old and bear twins or triplets. Some white-tails live as long as twenty years.

## The Truth about Deer

The deer population explosion sends deer bumbling into places where they don't belong. At least that's the human outlook. In Ashland, Kentucky, a deer jumped through a one hundred-year-old plate glass window that adorned a local art gallery. The animal ran through the gallery, smashing two artworks, and then escaped through a glass window on the other side of the gallery. Some witnesses thought the deer was being chased by dogs.

A high school science class in Iowa City, Iowa, was treated to a doe running around the classroom. The school's principal said, "Just out of the clear blue it crashed through the window. I've never seen a room evacuate so rapidly. The kids were diving out of the room."

A Davenport, Iowa, radio station was interrupted by a deer that crashed through the station's window. At the time WXLP was broadcasting a show called "Bulletins from the Boondocks." Guess so.

Just because you live in an apartment, doesn't mean you're safe. A Bloomington, Indiana, apartment complex was invaded by five deer, one of which crashed through a glass door and window. This deer then jumped over a wall into the complex's parking lot, landing on a pickup truck. Said one witness, "It sounded like someone was playing bumper cars in the parking lot."

Aircraft aren't immune either. A pilot crashed into a deer during a takeoff in Minneapolis, Minnesota. The private plane was demolished, and the deer was killed, but the pilot and passengers escaped unharmed. The airport manager said, "The pilot was at the wrong place at the wrong time. So was the deer."

In fact, deer will appear just about anywhere. Somehow, in 1990, a deer wandered into downtown Washington, DC. Nobody knows how the deer got there, but it disrupted morning rush-hour traffic as it panicked among the Pennsylvania Avenue cars. Exhausted and injured, it was finally darted, near some dumpsters at the Old Post Office Pavilion.

Some of these unhappy encounters result when a deer wanders into our territory and panics. But in other cases, deer wreck havoc when they find us intruding on *their* territory.

For example, during mating season, when a deer's hormones change, producing more testosterone in males, anything goes. In Caldwell, Texas, recently, an eight-point buck trampled a man to death. The man had been walking along a rural road, looking for antique bottles, which he collected. According to investigating biologists, the deer may have seen the man bent over, picking up bottles: Apparently to a white-tail buck, being bent over, ready to attack with antlers, is an aggressive posture. And so the buck behaved as any buck would.

Having spied another male in its territory during mating season, it attacked.

In another Texas incident, three surveyors working in the woods were attacked by a buck. One man was thrown twenty feet into the air; he landed in a creek.

One Texas Parks and Wildlife Department officer, who was chased by a buck, described the experience this way: "It's like having an ice pick running at you at thirty-five miles per hour, with one hundred pounds of force behind it."

Clearly, mating season is a time to avoid getting too close to these animals. (As is hunting season, but for a different reason—during hunting season you may be mistaken for a deer by a hunter.)

A sign of impending attack is when a buck pushes its ears backwards. A rigid body and stiff-legged walking are two other danger signs. Wildlife experts say if you are confronted by a buck, don't run—walk away. There's a good chance that the deer's natural instinct to escape will take over. Of course, if the deer has begun to charge you, by all means—run!

Yet, attacks on humans remain a rarity. Managers of parks, forests, and other natural areas say the biggest problem they face from deer comes not from the impact of deer on people, but from the impact of the burgeoning deer populations on natural resource areas.

Now, while it's nice to know this sort of background information, I don't feel that it captures the real truth about deer.

Let me proceed to give you the truth.

Let's start with what a deer is. Basically, there are five kinds: Bambi; deer that eat gardens; deer that carry ticks; deer that deliberately leap in front of cars; and deer that hunters strap to the hoods of cars.

## Bambi

By far the most common and popular form of deer is Bambi—the sweet gentle animal that you spy briefly, too quickly, in the woods. Children love Bambi. And so do most adults. After all, this is one of the most inoffensive, and cute, animals on the planet. True vegetarians and quieter than mice—how can anyone not like deer?

Witness the traffic backups in Shenandoah National Park whenever deer are near the road. Visitors are compelled to stop their cars and maybe get out to snap a picture or try to feed the animal. People like to see wild animals.

Bambi never hurts anyone, strolls majestically across the landscape, nibbles daintily upon a bough, and then disappears with a flash of white tail. Bambi poses for photographs. Bambi will almost let you touch his soft coat and gaze into his soulful, brown eyes. Deer who eat our apple crops, carry ticks, or dash in front of cars are Bambi's evil siblings.

## Deer Who Eat Gardens

You might not like deer if you have a garden. Jim Nolan, a nature writer, falls into this category.

In the past, "I never considered deer to be predators, because I never considered plants to be prey," he wrote. That was before he moved out of the city. Now he says, "This gardener views his turf as a writhing hotbed where hunter and hunted are constantly devising ingenious tricks to fool one another."*

The basic problem, as Nolan has discovered, is that if it's green, deer will eat it. That includes the items in your garden, as well as the tender shoots of young trees that you may have planted on your property.

Also, if it's brown, a deer will eat it, provided it's an acorn. In fact, for each one hundred pounds of weight, a deer consumes one and a half pounds of acorns daily. That's a lot of acorns, considering that a mature buck may weigh two to three hundred pounds.

A bountiful acorn crop can be a blessing and a curse for the gardener in deer country. A blessing, because the deer will forsake your roses, apple orchards, and other plantings for the acorns. A curse, because a large acorn crop one year means lots of deer the next year. Acorns don't fall throughout the year, so deer who eat them when they do fall will, at other times, be interested in other morsels your garden offers.

* Jim Nolan, *Spiritual Ecology: A Guide to Reconnecting with Nature* (New York: Bantam Books, 1990).

Nor is the problem limited just to gardens, as farmers around Cleveland, Ohio, will testify. There, large herds of white-tailed deer have been known to consume vast amounts of wheat. One farmer described the problem this way: "It's just like a bunch of cows were turned loose."

At Gettysburg National Battlefield Park, maintained to resemble the farmland that it was in 1863, there rages a terrible battle—against whitetails. Bob Davidson, a management assistant at the park says that seventeen years ago there were between four and five hundred deer on the park's four thousand acres. Today there are sixteen hundred deer. And what does the game commission estimate to be a suitable population? Eighty to a hundred.

The park leases land to farmers who maintain the agricultural appearance of the park, but in recent years deer damage has been so bad that farmers don't want to renew their leases. Gettysburg has started waiving fees if farmers can prove significant deer damage. The agricultural appearance is maintained at great cost: A historic peach orchard was restored only by completely fencing each young tree to protect it. Farmers have given up growing corn.

And there's overflow. "Right now we're very unpopular with the neighbors," says Bob Davidson. Deer are decimating landscapes and gardens all around the park. But the deer problem isn't unique to Gettysburg farmers. The Pennsylvania Farmer's Association claims 36.4 million dollars' worth of damage by deer each year.

## Deer That Carry Ticks

It used to be that the most common reason for wanting to keep deer away was gardens. Today, a more urgent reason is Lyme disease. Lyme disease, named after the Connecticut town in which it was discovered, is a bacterial infection, second to AIDS as the fastest growing infectious disease in the country (though not nearly as serious). You can get it when you are bitten by a northern deer tick, so named because it feasts on the blood of deer, but will also feed on pets and humans if given the opportunity. In the West, the western black-legged tick spreads Lyme disease but isn't as dependent on deer for food; it feeds on more than eighty different mammals.

The Lyme disease outbreaks are most serious in eight states: Connecticut, New York, New Jersey, Rhode Island, Delaware, Wisconsin, Maryland, and Pennsylvania. But smaller outbreaks do occur elsewhere. Fortunately, Lyme disease, if caught in its early stages, is treatable. If not caught early on, it can lead to chronic heart, tissue, and nerve disorders. (See Bugs, Slugs, and Scorpions for information on ticks.)

So now the white-tailed deer threaten your azaleas and your health. Estimates place the current deer population at twenty-five million— roughly the same number that roamed the country when Jamestown was a happening place. Of course by now there's much less territory for them and they're packed into small green spaces, often in populated areas.

At the Schuylkill Center, a private environmental education facility near Philadelphia, the deer population has increased twenty to twenty-five times beyond what the land can support. According to a *New York Times Sunday Magazine* article, the tick population at Schuylkill has increased along similar proportions.

## Deer That Leap in Front of Cars

Deer are also not popular with drivers. Car-deer accidents become common in many parts of the country during certain times of the year (usually fall and spring) when deer populations are high and the animals are likely to be on the move. In some cases, they're moving between areas looking for mates; sometimes they're looking for food. Roadside grasses are a favored fare; deer also lick de-icing salt applied to roads in winter.

In Pennsylvania, more than forty thousand deer died in car crashes in 1989. (This is not a misprint.) Indiana had more than eleven thousand that same year. Nationally, nearly a half-million deer are involved in these car-deer crashes, where the deer usually die and the car and driver are somewhat damaged.

Hitting a deer is usually more harmful to the deer than to the driver. Only about one hundred human deaths a year occur as a result of these crashes, according to the National Highway Traffic Safety Administration.

Even so, these run-ins can result in serious injuries to motorists and damage to cars. When you hit a buck and its antlers come crashing through your windshield, that's going to cause a lot of damage. And even if you don't die or get hurt when in a car-deer encounter, this particular peculiar combination of metal and flesh can be very costly. It's also very messy.

You might think these incidents would occur most often on high-speed interstates or limited-access four-lane roads. They don't, because fences keep the deer off the most dangerous sections of these roads. Instead, most of the accidents apparently occur on two-lane roads and four-lane roads with unlimited access.

The peak times for hitting a deer are between 5 A.M. and 7 A.M. and 5 P.M. and 8 P.M. Dawn and dusk are the times when deer are most likely to be active. These are also the times when a driver's visibility is reduced. Many deer accidents also happen at night, when the animals are temporarily blinded by a car's headlights: Deer may freeze in the road in front of the car, or even run right toward it.

The deer population keeps growing, so this problem is bound to get worse. One of the reasons for the explosive growth is that the populations of many of their natural enemies, such as bobcats, mountain lions, and wolves, have dwindled.

There's no birth control for free-ranging deer at present, except what nature provides: A harsh winter will cut down deer populations, but the United States has experienced unusually mild winters since the late 1970s. And too many deer trying to forage in any given area will eventually result in the animals stripping away available vegetation until there's nothing more to eat, causing starvation among the population.

## Deer That Are Strapped to Cars

Hunting groups like to offer hunting as the "humane" alternative to starvation when populations of deer are at their peak. While there is considerable controversy over whether hunting is, in fact, humane, safe, and ethical, there is no question that it is popular. (Ann Landers received twenty thousand letters in response to a column she ran on illegal deer hunting.)

In spite of the all the controversy surrounding it, hunting has little effect on population size; it neither stabilizes nor reduces the population size. The number of hunting licenses sold has dropped steadily since the early 1980s. In 1989 there were about a half million fewer licenses sold than in 1981. No matter how many hunters there are, it would take an unmitigated act of carnage to significantly reduce the deer population. As long as the habitat we provide is so attractive, we'll have deer.

With all the deer around, you'd think it would be pretty easy to get a deer, but it's not. Only one in four hunters bags a deer. Hunters generally oppose any efforts to reduce the deer population. As a matter of fact, they often join hunt clubs where they have access to private land where the deer are treated to corn and salt licks. Since hunters traditionally go after large bucks with trophy-quality antlers, does tend to survive the hunting season and continue to reproduce.

If deer are your problem, however, I have the answer. In fact, I have several answers, designed to protect your garden and car.

## Garden Protection

In the absence of meaningful predators, deer thrive. But, to survive, the animals eat everything in sight. Deer don't present a problem for mature trees, but are deadly to young trees that are just beginning to break through the ground. By stripping an area of its vegetation, deer also deprive other creatures of food.

Evidence of deer overpopulation: a browse line where deer have consumed every leaf, blade, needle, flower, and twig from the ground up to a height of six feet. They eat just about anything, especially when they're really hungry, and can digest over six hundred different species of plants.

It's pretty easy to determine when a deer is the one robbing your garden. They leave obvious signs: tracks and droppings. Without these signs, assume that the problem is a rabbit or other rodent, or a neighbor who's too lazy to grow his own vegetables. Another indicator is unevenly torn leaves; deer eat by tearing because they do not have upper incisors. Deer also feed about four to six feet above ground, well above the reach of a rabbit. How do you keep deer out of your garden?

The only solution that's perfect is to buy all your vegetables at the supermarket. Easy for a city slicker to say.

Alternately, focus on the garden's perimeter. Electric fences are a pretty good deer deterrent, although occasionally deer will bolt the fence. Also, sometimes a broken wire stops the current, making constant vigilance necessary.

Electric fences also keep other animals away, because the animals will become conditioned to avoid the fence. A couple of encounters with an electric fence and the deer will learn to avoid it even when it's not activated. You'll have to reelectrify the fence when new deer come into the vicinity, but you probably won't have to keep the fence juiced up, wasting valuable electron-cash units—otherwise known as money.

For smaller enclosures (forty by sixty feet), snow fencing works well. Enclosing larger areas is usually too much for the pocketbook. Your local agricultural extension service can give you instructions for several types of deer fences.

Instead of fences, you might opt for thorny bushes. The higher the better, but even a few feet tall should work. While deer can leap over a fence as tall as seven or eight feet, once a deer miscalculates the height of a bush full of thorns, it will think twice about nibbling in your garden. Thorny trees work well, too. Depending on what grows where you live you might select such plants as Russian olive, hawthorn, or Japanese barberry. There are plants deer don't like to eat, and maybe you can consider deer-proofing your landscape. Try holly (the pricklier, the better), barberry, boxwood, English ivy, daffodils, irises (if you can keep insect pests out!), spruce, fir, and prickly plants like cactus. Strangely enough, deer will eat roses, despite the thorns.

At the opposite extreme, fruit trees, tulips, strawberry plants, and day lilies attract deer. They're also especially fond of azaleas, arborvitae, and yew.

Still another way to keep deer off your property is to get a dog. A small one won't do; after a couple of encounters, the deer will figure out that the dog is just a yapper. Big dogs seem to do a pretty good job of discouraging these intruders.

Or, you can dissuade deer from browsing on your boughs by using one of the many deer repellents on the market. There are contact repellents that taste bad and area repellents that smell awful. It's best to apply the contact repellents when the plants are dormant; then animals don't feed on the buds in the winter. Naturally, you have to respray the new growth. No repellent is good for large areas, but they'll work well in small gardens and yards. While repellents won't eliminate damage, they should lessen it, and the contact repellents are generally more effective than the area ones.

To use a repellent, you have to be vigilant. Repellents weather, so they have to be reapplied often, and you have to watch for any new growth that needs treatment. Plus you have to remember that deer can reach up to six feet, so you have to treat high and low.

Some of the better-known repellents include: Big Game Repellent, Deer Away, Magic Circle, Hinder, Ropel, Chaperone, Gustafson 42-S, and Nott Chew-Nott. These can be purchased at garden supply stores or through mail-order gardening catalogs. (See the resources section for some suppliers.)

Many people rely on rather grisly, smelly, old-fashioned remedies: human hair, feather meal, blood meal, or meat meal. Human hair sometimes works because deer are afraid of people and first notice humans by smell. Try human hair from a barber or hair salon stuffed into stockings or mesh bags and hung in trees or along fences. Be sure to change the hair often. Feather meal is pretty hard to come by, but you can get it from a poultry processing plant. Tankage, putrefied meat scraps from the slaughterhouse, and blood meal are other options. You can put the tankage in baggies or punctured cans and hang it in trees or around target vegetation. Carnivores may pull the smelly stuff down.

To keep deer from nibbling on the leaves of trees, some people even hang bars of deodorant soap from the branches. The scent of Irish Spring is supposed to be especially effective, according to deer-plagued residents around Philadelphia.

Many folks swear by the old standby repellent of mothballs or flakes sprinkled around on the ground or hung in mesh bags. Still, don't forget that naphthalene, the active ingredient of mothballs and flakes, is flammable. And its noxious effects aren't limited to nonhumans. It affects us too: Prolonged inhalation can bring on headache, nausea, and vomiting. In fact, there are reported infant deaths from dermal absorption of naphthalene in blankets.

Most of the commercial contact concoctions taste awful and discourage animals from eating your sprayed plants. Unfortunately, deer have to taste it once to know how rotten it is, so there will be some loss of leaf. The drawback: Sprays taste as bad to humans as they do to animals, so you can't use them on any food crops. And these potions are designed to soak into the plant and render it inedible, so don't count on being able to wash it off.

Deer Away is an area repellent, also sold under the name Big Game Repellent, manufactured from putrescent egg solids. The smell, which is mild to humans, is overwhelmingly awful to deer, and they won't even taste the first leaf before they decide to find dinner elsewhere. As with other sprays, you won't want to eat anything sprayed with Deer Away, which comes in powdered and liquid form. The product was developed by Weyerhaeuser to protect their seedlings from foraging deer.

Anxious to protect the state's agricultural businesses, in 1980 the Pennsylvania Cooperative Fish and Wildlife Research Unit conducted a study of fourteen repellents to find the most effective ones. The final report stated, "Our tests have shown only one repellent, Big Game Repellent, to be significantly different from no treatment at all. . . . This does not mean that other repellents will not work in a given damage situation. We do know, however, that repellents vary in effectiveness. Some deer may be discouraged, but others may not; variable deer numbers and feeding pressure are factors to be considered."

Working with captive deer, the Pennsylvania researchers tested the "other" effective repellents mentioned above. Along with Big Game Repellent, these included meat meal, feather meal, Hinder, hot sauce (about two tablespoons to a gallon of water sprayed on leaves), Chew-Nott, Flowable Fungicide, and Gustafson 42-S.

One spray repellent that was not reviewed by the Pennsylvania researchers, but which has received EPA approval for use on food crops, is called Hinder Deer and Rabbit Repellent. The spray's smell repels deer and rabbits and will work for up to eight weeks, although it's best to respray areas after three or four weeks. Hinder's active ingredient is ammonium soap. It works best if sprayed as a border on the grass surrounding a protected area so the animals will learn to avoid the sprayed plants.

Niles Kinerk, director of the Gardening Research Center at Gar-

dens Alive! an Indiana organic product business, tests the deer repellents his firm sells. He says, "Hinder is about as effective as Deer Away,
but the nice thing about Hinder is that it has registration for use on
food crops. The trick with repellents is to get one that works at a high
rate, say ninety percent, and the other thing is to use it properly, get
it on before the deer have begun to browse on the plants. You especially need to start spraying in the early spring when everything begins
to grow."

However, he adds, nothing will be totally effective. "If you've got
a starving animal and they don't have any other options, they'll still eat
the sprayed plant."

Mark Fenton, of California's Peaceful Gardens, a supplier of organic garden products, backs him up. "No deer repellent is foolproof.
The only foolproof deer repellent is an eight-foot-high fence."

However, Fenton offers some suggestions on ways to get the most
out of any repellents that you use. "The thing you want to do with deer
repellents is to rotate the different types frequently. You don't want to
use one for two months and figure the deer is not going to get used to
it." In fact, the hungrier the animal is, the more audacious it will
become. They will keep testing, and eventually they will pass through
the repellent and lose some of the fear of it, he says. "That's why, if you
keep using new scents, they don't quite get the ability to keep pushing
it, because they don't get used to one."

He says the scent that lasts the longest of all the repellents his firm
carries is called Magic Circle. It's a bone-tar oil, "and it's pretty smelly,
it has a strong odor." Might even keep hunters away. Another one that
Fenton recommends is produced from a lion urine extract, but it
doesn't last as long. Neither of these can be used on food crops.

You need to be relatively selective in applying area repellents: Put
them where you have seen deer or where you would prefer not to see
deer. The alternative is to have your entire yard smell ghastly.

Skittish animals, deer are also easily frightened by noise or movement. A plank hung from a tree limb in such a way that it bangs
against the tree when the wind blows can frighten deer away. So, too,
can wind chimes and radios that are left on. Sheets of aluminum foil
or mirrors may also bother deer enough to keep them from your
vegetation.

The loud detonations from gas exploders may frighten away deer as well. You can set a timing device to detonate at regular intervals, but you'll need to vary the interval from day to day so the deer don't get used to it. And to keep things really interesting, you should move the machine every few days or so. The devices sell from about $150. As with repellents, deer can get used to loud noises, too; they have been known to live near airports and shooting ranges. Firecrackers and shotgun blasts may do the job temporarily.

Outwitting deer requires experimentation: Deer are nervous animals—you won't know what will keep them away until you try.

## Deer in the Home

They're never invited. They barge right in.

In 1990, an Omaha, Nebraska, house was invaded by a buck that crashed through a window, rammed a personal computer, tore down drapes, and otherwise ruined the inside of the house. The deer had been frightened by a passing car and had run through the nearest opening—a window.

A deer ran around an Edmonton, Canada, house for a half hour before running out the front door. The deer, which had entered the house by breaking through the screen door "looked like an out-of-control Doberman," according to the frightened homeowner, who had spent most of his time trying to prevent the deer from running up the stairs.

There's nothing you can do to discourage this type of freak occurrence, but if Fate sends a deer your way, just block off what rooms you can, open the door, and get out of the way.

## Preventing Car Accidents

Before you develop a fear of deer (bambiophobia), let me add that the chance of hitting a deer is tiny for any given individual. That is, the deer accident rate per mile traveled is small.

But if you're worried about being the One, read on.

To avoid an unwanted run-in, drive with both your eyes and mind open. That means:

• Look out for deer during the times of day when they're most likely to be active (dawn and dusk).

• Look out for deer when you're driving through areas that are likely to house deer (forest edges) or that might contain animal travel lanes (for example, a path or dirt road coming out of a forest).

• Drive more slowly when you're in deer areas. The likelihood of injuries increases for drivers who drive fast (and for drivers in smaller, lighter vehicles).

If you see a deer on the road, slow down as much as you can. Swerve to miss the deer, if you're sure you won't end up hitting a tree or another car. But be forewarned: When one deer has crossed a road, there's probably at least one other deer behind it.

To make roads safer, highway departments in several states are experimenting with a variety of devices. One technique that seems to work is putting reflectors along roads that deer frequently cross. When hit by a car's headlights, these reflectors create the illusion of a fence, thwarting a deer's approach. Some states have dug underpasses to give deer a safe way to cross roads. Still others have erected off-the-road deer feeders to keep the animals away from the roads in the first place.

Another tactic is to mount a "deer whistle" on your car, but there is a difference of opinion as to whether these devices really work. The whistles, which cost about $10 and can be bought in catalogs and in many automobile parts stores, are installed on the car's front grill or bumper. As the car moves along the road, the whistle emits a shrill, ultrasonic cry that warns deer away, but which can't be heard by humans. The whistle can be heard up to 1,200 feet away.

To reduce insurance claims in Indiana, one Indiana insurance company offered deer whistles for sale to its policyholders. The company sold fifty thousand whistles the first year, and according to a company official, "We saw a substantial decrease in the number of deer claims."

But a biologist disagrees. "Deer whistles are a joke," he says. Nobody's ever done a scientific study to show that they work, and nobody's studied whether deer can actually hear the whistles, he says. The only studies that have been done, he asserts, are marketing studies—which turned out "okay."

The California Highway Patrol takes a middle ground, claiming that deer can only hear the whistle when a car is heading directly toward the deer. Otherwise, the sound doesn't reach the deer's ears.

## ❖ DEBARRING DEER

1. Try different repellents to keep deer from feasting on your landscaping and kitchen garden.

2. Get a dog.

3. Fence areas you want to protect.

4. Feed the deer something more attractive than your precious plants—corn, for instance.

# DIGGERS: MOLES AND COMPANY

## MOLES

You might say that moles have tunnel vision. They spend most of their time underground hanging out in tunnels of their own construction. There they mate, raise families, eat, fight, and sleep. In fact, one of the few reasons they go above ground is to find a new and better place to—you guessed it—build tunnels!

Nature designed the mole with this mission in mind. Moles are small mammals, between four and six inches long, with large claws and noses. In fact you'll probably identify the mole by its large, pink probiscus, a tender appendage they would never, ever use as a digging tool. They use the long, strong claws to dig. Their eyes and ears (well, they don't have external ears) are hardly noticeable.

There are seven species of moles in North America, although the most common ones in the United States are the eastern mole and the star-nosed mole. While both are found in the eastern half of the country, the eastern mole is especially prevalent in the Midwestern states, the East and Southeast, while the star-nosed mole is found more often around the Great Lakes. What all moles have in common is tunneling; they rarely come to the surface and live quiet, solitary lives underground. Moles may share runway networks, especially if their home ranges overlap, but they don't socialize—in fact, they will fight when they encounter one another.

They tunnel underground because that's where the food is. They eat a lot and so must travel far and wide to get all the food they need: A captive mole will starve in a few hours unless fed nourishing food.

Moles produce two types of tunnels: tunnels close to the surface that we see (and, if we're mean, step on) and tunnels deep underground.

Moles dig the surface tunnels when they're hunting insects and worms. The tunnels often connect to deeper tunnels that lead to home chambers. Moles make surface tunnels by pushing the soil up to form the roof, rather than pushing it out the end of the tunnel. During wet weather, surface tunnels are very shallow, but when it's dry, the moles tunnel deeper for food.

Deep tunnels may be six to twenty-four inches below the surface. With so much heavy earth above these tunnels, it's impossible for the moles to create the tunnels by pushing the "ceiling" up. So, after loosening the dirt, the moles push it back with their hind feet through the completed section and back to the surface or into an old shallow runway. As the mole makes the tunnel longer, it creates new exits so it doesn't have to push the excess dirt quite so far. Moles keep nests in these deeper, safer, tunnels for resting and raising families. The homes are in high and dry spots, perhaps under buildings or sidewalks. Tunnel systems may be used by successive generations of moles, and a mole continues to use the resources it creates. They have definite home ranges that may overlap.

Moles eat insects (and insect larvae), grubs, earthworms, carrion, the unwary mouse (rodents sometimes use mole tunnels as a shortcut), and maybe even other moles that they find in their tunnels. When food underground is scarce, they'll hunt above ground, catching frogs and small mice. A mole eats between seventy and a hundred percent of its body weight each day.

Moles thrive on our manicured lawns and golf courses, especially shaded areas. They don't live in dry, semiarid zones. Larger mammals like coyotes, dogs, badgers, and skunks will dig out moles, and other animals may eat a mole that ventures above ground.

Moles mate about once a year. The female raises the young moles in her den, which is lined with grasses and other material gathered above ground. She may prepare several nests, some of which may be above ground under a log or in a pile of vegetation.

After a gestation period of four to six weeks, usually four young are born. They are helpless, furless and blind, at birth, but are nearly adult-sized and sighted by three weeks. After about seven weeks they are on their own and must find their own home ranges. Often they must travel above ground, one of the few times they'll be found there, until they find a spot. They live about three years in the wild.

## Man and the Mole

Golfers don't appreciate moles adding extra holes to their already difficult courses, so I guess they have a legitimate gripe against moles.

Many homeowners—at least the ones that prize level, manicured lawns—don't like moles either. Or think they don't. In fact, because moles eat insects like Japanese beetles and grubs, and aerate the soil during their tunneling, they shouldn't be considered a pest. They carry humus deep into the soil, and bring subsoil close to the surface. Moles are one of the most abundant of small mammals. Their innocuous behavior combined with their positive ecological role, make them a boon—if you can overlook those little rises in your lawn.

Homeowners also often falsely accuse moles of causing damage to flowers, gardens, and landscape plants. Moles don't particularly like plants; any plant damage is more likely the work of voracious vegetarians, like voles, or perhaps mice. Niles Kinerk, director of the Gardens Alive! Natural Gardening Research Center in Indiana says, "I don't mind them myself. Moles are kind of beneficial."

But he understands how some people might think otherwise.

"You run into these people who spend $700 a year to have their lawn fertilized and manicured and have special lawn mowers that cut it in a special direction. Some people are pretty obsessive about their lawns," he says.

There are several things you can do to encourage moles to do their tunneling elsewhere, Kinerk says. You can apply milky spore or beneficial nematodes to your lawn; they kill the grubs, a main food source for moles. However, the problem may get worse before it gets better, as the hungry moles frantically search the area for food before heading to greener pastures.

Another approach for a relatively small area is simply to water it thoroughly; getting the ground wet one or twice a week for three or four weeks may deter them, says Kinerk. Alternatively, you can try drying out the soil and forcing the insectivorous mammals elsewhere. Packing the soil with a roller may make it too compact for digging.

Repellents may work. Mothballs and flakes of course may discourage the diggers. An old folk remedy, planting the castor bean or caper spurge may repel the animals too. Remember, though, other animals and children may try to eat the mothballs you spread in your yard or, even better, drop down a mole hole. Electromagnetic repellers? Save your money.

Another solution is the stink-em-out routine: Sprinkle a castor oil concoction on the ground. To make, whip two ounces of castor oil with one ounce of liquid dish detergent in a blender until it holds its shape. Add water equal to the volume in the blender and whip again. Fill a sprinkling can with water and add two tablespoons of the castor mixture and sprinkle on areas of heaviest concentration of burrowing. This works best after a heavy rain or watering.

Vibrations can also scare away moles. Try pushing a child's pinwheel down in the earth in several spots in the area where the tunnels are dug. Or you can purchase a commercial mole windmill or electric vibrator like Plow and Hearth's Mole Evictor. Each device is effective for five hundred square yards, and you can choose between solar-powered or battery-powered machines. Unlike the children's pinwheels, these work without wind and send out mole-annoying sound waves from their buried aluminum shaft.

Another approach—one your kids will probably enjoy—is to walk on the tunnels and collapse them. While this sounds cruel, you're unlikely to actually kill the moles, who live in the deeper tunnels. But if you stomp the tunnels frequently enough, the mole will get the idea that the area is unsuitable and abandon it.

Trapping is probably the most successful approach. An infestation of moles would probably be three or four an acre, so you should be able to catch the animals. Some mole traps impale unsuspecting tunnelers as they go about the business of getting a little dinner; others cut them in half or choke them. And if there are more moles nearby, they may move into the open habitat of the recently deceased. It's not a pretty business.

But don't despair! You can live-catch moles. Find an active mole runway and set a pit trap. Dig into the runway and place a three-pound coffee can or wide-mouth quart jar in the dirt below the runway. Then cover your excavation with a board, making sure no light gets in. And wait (while checking twice daily). But release your captive mole far away from your lawn.

### ❖ MARSHALING MOLES

1. Treat your lawn for grubs, a main food source.
2. Pack the soil down or water thoroughly to make it difficult for the moles to dig.
3. Actively harass the moles so they'll seek housing elsewhere: stamp down runways and use mothball and plant repellents.
4. Trap the moles alive.

## POCKET GOPHERS

In function, but not in family, pocket gophers (so-called because of the fur-lined pouches for carrying food next to their mouths) are a close cousin to moles. Both really dig tunneling.

But while moles are beneficial, many people find gophers downright destructive. Larger than moles, they create bigger holes—making them unpopular with golf course owners. Bulb and root-eating vegetarians, they are the nemesis of gardeners and farmers alike. And with large incisor teeth that can cut through almost anything in their path, they've earned their share of enemies in other quarters.

For example, gophers managed to raise the ire (not to mention the roof) of at least one family in Minnesota. It seems that in 1989, a gopher chewed through a thick polyethylene gas pipe leading into a home in St. Cloud. Apparently the gas accumulated and ignited, blowing the house up.

Gophers even can chew through metal, including irrigation lines. In fact, in Nebraska, pocket gophers are said to reduce alfalfa crop yields by twenty-five percent. The losses are due to the gophers' severing of electrical and irrigation lines that happen to be in the path of their tunnels, and by their eating of alfalfa roots.

The gopher is a small mammal, six to twelve inches long and weighing from one-half to a full pound. There are thirty-three different species, but the gopher is found in some form or another throughout the United States.

Gophers have long claws on their forepaws for digging and strong shoulders. Because they spend most of their time in the tunnels, their long whiskers help them sense their way around the darkness, and their tails help guide them when they go into reverse. Like the beaver, their lips close behind their large incisors.

The gopher rarely ventures above ground, except perhaps to gather bark, grass, forbs, or greens, which the animal stuffs into its cheek pouches for later feeding underground. They eat roots and also pull vegetation into the tunnel from below. Depending on the species, the pocket gopher prefers different foods. The gopher's main life is a subterranean one. An efficient digger, a single gopher can tunnel two or three hundred feet in a night, and tunnel systems may be as long as two hundred yards. All this dirt ends up in mounds (about 150 mounds per year per gopher) piled beside their holes.

Why all this emphasis on earth moving? Gophers tunnel beneath the surface of the soil, eating the roots and tubers they encounter in the process. Unlike moles, they don't tunnel close to the surface and push up the soil. Gophers also use their tunnels for breeding, nesting, and resting places. Since gophers don't hibernate, tunnels also have to be deep and warm, although some northern-living gophers will tunnel in the snow and pack the tunnels with earth. When the snow melts, the earth tubes remain above ground until they disintegrate. Burrow systems may be straight tunnels or branched labyrinths. How far the gopher tunnels depends on how much food is in the area. One gopher lives in each tunnel, unless mating or caring for young.

In the northern parts of their range gophers breed once a year, but in the irrigated fields of California, they may breed year-round. The female gives birth to from one to seven young, and in some species, the male may stay around long enough to help care for the young. Once the young have weaned, they leave and begin to dig their own burrows. They're sexually mature at a little under a year and live one to three years.

Pocket gophers have plenty of enemies. Weasels, skunks, and snakes come into the tunnels to find them. Other predators dig them out—badgers, for instance. Above ground, a world of predators— raptors, coyotes, fox, house cats, skunks, and bobcats—wait for a gopher to deposit a little soil on the mound or venture out for a bite of grass.

## Going After Gophers

Don't confuse chipmunks with pocket gophers. Chipmunks are harmless little guys that scurry above ground and feed on seeds and nuts. Although gophers churn and aerate the soil, if a gopher is in your garden, it is not for altruistic purposes. Gopher's in your garden are after plants, and they are especially fond of bulbs.

One effective, if labor-intensive, way to protect your plants is with root guards, according to Mark Fenton of Peaceful Gardens, an organic garden supplier in California. Root guards—which you can make or buy ready-made—form a protective cage around the plant's roots. Fenton says you can use chicken wire to make your own root-guard baskets. However, small gophers can get through even one-inch or three-quarter-inch holes. So, if you want greater protection afforded by half-inch or smaller holes, use aviary wire.

Some people use root guards for flower bulbs; others for ornamental or fruit trees, he says. Obviously it's much too labor-intensive for vegetable gardens. Tree roots will be able to get through the wire as the tree grows, although this will eventually cause the guard to become ineffective.

An easier solution, however, is to trap the gophers. Like most mole traps, gopher traps are deadly. Gopher tunnels are a little more difficult to find than the surface mole tunnels: You have to dig a ring about a foot and a half away from the mound until you find the main tunnel. You could probably live-trap them in a large pit trap, catching them in a gallon-sized jar. Or, you could try getting a large cat. "There are a lot of cats that are good gopher cats, and they will keep yards clean," says Fenton. "It probably depends on how well they are fed."

You can also try keeping the soil very wet. It's difficult for gophers to dig in the wet earth; it clumps on their claws and gets their fur dirty. In addition, wet soil will trap noxious gases in the tunnels with the gophers; the dry soil naturally allows air exchange.

Many companies sell earth vibrators, stakes that when implanted in the ground send out small shock waves and frighten away burrowing mammals. As with other gadgets sold for garden and yard use, it's anybody's guess how well (or even why) they work. Fenton says that manufacturers claim the effectiveness of the device depends on where it's placed and on the structure of the soil (which affects how well vibrations travel through it). "I think it works not as well in some sandy soils, because the vibrations don't travel as far in sandy soils as they do in a heavier clay soil."

Some people poison gophers. In fact, Nebraska's alfalfa farmers are so incensed about pocket gophers, that many pour poison down gopher holes. Some people have tried mothballs or flakes as a repellent. Fortunately, technology may be coming to the rescue in this case. Researchers are breeding alfalfa plants that not only seem resistant to gopher teeth, but that are actually helped by gnawing. As a matter of fact, horticulturists are designing gopher-resistant plants even as you read.

Maybe then science can develop a tulip bulb that's gopher-friendly.

## ❖ GETTING GOPHERS

1. Use root guards to protect bulbs.

2. Scare gophers away with earth vibrators.

3. Water the soil to discourage gophers' digging.

4. Get a cat (from your local pound or animal welfare society!).

# WOODCHUCKS

One question that people frequently ask about woodchucks is: "How much wood could a woodchuck chuck if a woodchuck could chuck wood?" In fact, since woodchucks don't chuck wood, and probably wouldn't if they could, this is a ridiculous question.

It should be rephrased as follows: "How much earth could a woodchuck move if a woodchuck could move earth?" This doesn't, admittedly, have the same ring to it. However, since woodchucks can move earth, it is possible to provide an answer. According to the U.S. Department of Agriculture Extension Service, a single woodchuck can move more than seven hundred pounds of earth in one day. Whether they do or not, however, is anybody's guess.

Woodchucks are found in the Plains states and throughout much of the eastern United States and southern Canada. Also known as groundhogs, these animals have compact, chunky bodies sixteen to twenty inches long covered with a grizzled brownish gray fur and supported by short, strong legs. Males weigh five to fourteen pounds; females are a little more petite.

The scientific name for woodchucks is *Marmot monax*. *Marmot* places them in the squirrel family; *monax* is derived from the American Indian word for digger. And, true to their name, woodchucks are diggers. They burrow into the earth to create their homes—tunnel and den systems where they hide when threatened, mate, wean their young, and, in winter, hibernate.

Woodchucks are specially equipped to do this dirty work. To dig up the earth, they have short, powerful legs equipped with strong claws. The woodchuck pushes loosened earth out of the burrow with its blunt head and chest. To keep dirt out of their ears—a problem if you happen to be an animal that pushes loose dirt around with your head—these earthmovers can actually close their ears.

Woodchucks prefer to locate near open farmland. Their burrows commonly are found in fields, pastures, along fence rows, stone walls, roadsides, at the base of a tree, or adjacent to a garden. You can find active burrows in spring or summer by looking for a large mound of freshly excavated earth at the main entrance.

The main entrance is ten to twelve inches in diameter. In addition, there are usually between one and four escape holes. These secondary holes are harder to find (for humans, that is), because they're usually located in thick vegetation; also, because they're dug from below the ground, they do not have mounds of earth beside them.

The length of a tunnel varies from eight to sixty-six feet, and leads eventually to a nesting chamber, which is about sixteen inches wide and twelve inches high. Woodchucks use the den for several seasons and hibernate in it during the winter.

Woodchucks spend so much energy on their tunnels for security reasons: Woodchucks have a number of enemies—including hawks, owls, foxes, bobcats, weasels, dogs, and man—and they don't run very fast, so they like to have a hole nearby to drop into if things get tense. In fact, a woodchuck usually ventures no further than fifty to one hundred feet from its den in a typical day, although the distance may vary based on the availability of food.

When a woodchuck wants to see if it's safe to venture out, he pokes the top of his head out over the rim of the burrow. Interestingly, woodchucks' eyes, ears, and nose are located toward the top of their head. This adaptation allows them to see if the coast is clear, while keeping most of their body hidden in the tunnel.

The main reason woodchucks leave the burrow is to eat. Woodchucks are primarily vegetarians. Like other rodents, woodchucks have white, chisel-like incisor teeth which they use to efficiently chomp away at the products of our fields and gardens. They feed above ground, mostly in the morning and evening. Woodchucks feed on a variety of wild grasses and field crops, like alfalfa, clover, and legumes. They also like vegetables—including beans, peas, and carrot tops. The more tender and succulent, the better.

When not feeding, woodchucks sometimes spend the warmer part of the day basking in the sun. You can sometimes spot a woodchuck dozing on a fence post, stone wall, large rock, or fallen log. These periods of leisure, however, are usually spent close to the burrow entrance. If danger threatens, the woodchuck will scurry inside.

When a woodchuck is startled, you might hear it emit a shrill whistle, or alarm. This is followed by a low, rapid warble that sounds like "tchuck, tchuck"—the sound that inspired their name.

As colder weather rolls around, woodchucks go into hibernation. In fact, they are among the few mammals that enter into true hibernation. This usually occurs in late fall, near the end of October or early November, although the start of hibernation varies with the latitude. They continue to hibernate until late February or March, when they will emerge from their burrows to find a mate.

Breeding occurs in March and April. After a gestation period of about a month, the female gives birth to a single litter of two to six (usually four) young. The babies are weaned by late June or early July, and soon thereafter strike out on their own. They frequently occupy abandoned dens.

Mature woodchucks will often use a burrow and den system for several seasons. They keep the burrows clean, and annually replace nest materials. When they abandon their homes, other animals—including rabbits, skunks, foxes, and weasels—may move in.

One benefit of all of this burrowing is that it actually improves the soil. Woodchucks condition the soil through their burrowing activities. However, it is hard to convince farmers and gardeners that these are benign creatures. The problem is that they like to eat some of the same foods we like to grow. They love gardens—and they're notorious for climbing fences to get into one.

## ❖ CHUCKING WOODCHUCKS

**1.** Make your garden harder to get into. Fencing can help reduce woodchuck damage. However, woodchucks are good climbers and can easily scale wire fences. If you decide to fence the area, bury the lower edge ten to twelve inches below the surface to prevent woodchucks from burrowing under the fence. The rest of the fence should extend three to four feet above the ground. Then, to keep woodchucks from climbing the fence, install an electric hot-shot wire four to five inches off the ground and the same distance outside the fence. (Some people have found that a hot-shot wire alone—without a fence—will deter woodchucks from entering a garden.) Before installing the hot wire, remove any vegetation near it so the system won't short out.

Another way to keep unwanted burrowing animals out of your garden is to plant only things they don't like. Scilla and castor bean discourage gophers and moles, but castor bean is poisonous so watch it when children are around. Try different vegetables—eggplant, raddicio, or broccoli. Maybe they won't like it.

**2.** Chase the woodchucks out of your area. There are several different ways to do this. Because woodchucks are easily frightened, you might try installing a large leashed dog. This will probably keep them away, although then you have a dog to care for.

Another approach involves finding woodchucks' burrows, then making life unpleasant for the residents so that they'll move of their own accord. You can determine which tunnels they're using by stomping down any mounded openings you find; the ones that are re-dug are active.

Then, you drop unpleasant things into a tunnel opening. Various authorities recommend shoveling in some dog droppings or stuffing up the hole with a rag soaked in peanut or olive oil. (Apparently the oil becomes rancid and will stink the animals out.)

**3.** For urgent situations, there are harsher solutions.

One of the more common means of woodchuck control is the use of a commercial gas cartridge. This is a specially designed cardboard cylinder filled with slow-burning chemicals. It is ignited and placed in the burrow system, and all entrances are then sealed. As the gas cartridge burns it produces carbon monoxide, which accumulates in the burrow system and kills the woodchuck. Gas cartridges are available from local farm supply stores and the U.S. Fish and Wildlife Service.

You might also consider traps. If you use live traps, check them twice a day so the animals don't suffer while waiting transport. Release them where they'll find plenty of wild food, like grasses. Don't trap them in the late summer, before they hibernate. Wait until the early summer, after they've awakened from hibernation and are well-fed and strong.

You might also try shooting unwanted woodchucks; apparently their meat is tasty. In some states woodchucks are game animals; in other states they are not governed by any hunting regulations. Check your state's regulations before you take the .22 out in the backyard.

# MICE AND RATS

## MICE

Our society has a rather schizophrenic attitude toward mice. Consider the fact that in books and cartoons, mice are celebrated: There's Mickey Mouse, Minnie Mouse, and Mighty Mouse and other pint-sized squeaky heroes. When it's cat-versus-mouse in the cartoons, the mouse is always the good guy. And, witness the many varieties of cute, stuffed mouse toys for children.

So naturally, as a child, I always wanted to have a real mouse as a pet. I could never figure out why my parents wouldn't allow it. Until I grew up and became a property owner, and suddenly recognized that mice were, in fact, the bad guys: Something to be gotten rid of.

Why? They like to eat your food before you do. Mice enjoy tearing apart pillows, mattresses, stacks of magazines, and other shredable objects to turn into nests. Mice leave droppings all over the place, too.

And, they carry disease, including rat-bite fever, tapeworm, ringworm, leptospirosis, and salmonellosis. Leptospirosis, which is spread through urine and can cause kidney and liver damage, is rare, and rat-bite fever is even rarer. Salmonella or other microbial food poisoning is a greater possibility, with probably thousands of cases each year. You can get these diseases from mouse bites or by eating food that mice have been munching on—and contaminating through their poor table manners.

And they're destructive. Maine resident Jane Connors discovered mice living in her couch. Initially attracted by food crumbs dropped by snacking TV fans, the mice decided to stay. They gnawed and burrowed their way into the furniture and made a nice nest of accumulated foam rubber and fabric. Connors only discovered the hole when she was looking for the TV remote.

At the outset I have to tell you that getting rid of mice won't be easy. Mice *like* living with people. They thrive in dwellings that humans inhabit. They can live on as little as one tenth of an ounce of food a day (something in the neighborhood of a few Cheerios), and can get their water from almost any place there are a few drops.

There are more than 250 kinds of mice in North America. The white-footed mouse and deer mouse are the most common ones you'll see outside, though they'll come indoors, too, seeking food and warmth. Other species, the brush mouse, cotton mouse, and pinon mouse, for example, are more specialized and limited in range. The familiar visitor to our domiciles, the house mouse, is a different species and was introduced to the continent in the sixteenth century. Outside of rural areas, it's most likely the mouse you'll see or hear in your house. It's originally from Asia. You can tell them from white-footed and deer mice, because house mice are gray and have long, scaly tails, and the white-footed mice are brown to near black with white undersides and feet.

Despite the name, house mice may live outside of human homes, even in open fields. But they're most likely to be in the vicinity of homes and commercial buildings. I'm going to include white-foot and deer mice in this section, because the general biology is similar. Mice have small home ranges, and may only travel within a small woodland, barn, or home. Studies show that they usually settle within five hundred feet of their birthplaces. All mice enjoy a visit inside your house, especially when it's cold, and some stay too long.

House mice eat nuts, seeds, fruits, berries, mushrooms, and insects—or whatever is in the cabinet. White-footed and deer mice will also eat fruit, insects, fungi, and possibly some green vegetation. Even house mice prefer seeds and grain and are dedicated nibblers, trying everything just to see if they like it. And they probably will. Vulnerable pantry items include high fat treats like nuts and chocolate candies. It's not that they eat so much, just about three ounces a day, but they taste everything. They destroy and damage far more food than they consume. House mice can get by on little or no free water because they get the water they need from the food they eat.

Since they don't hibernate, some house mice cache food for the winter, and they hoard more food in colder parts of the country than they do in more temperate climates. White-footed and deer mice will hoard food near their nest sites, especially in fall and winter. They don't necessarily eat the food they hoard, and the attractive caches call out to a host of other pests.

Mice are generally nocturnal, but you may see a house mouse during the daytime. While mice have good hearing and senses of taste, smell, and touch, they don't see very well and are color blind. They're good athletes and can generally do whatever gnawing, climbing, jumping, or swimming it takes to get where they're going.

A house mouse doesn't go far daily, probably traveling an area ten to thirty feet in diameter. They're true creatures of habit and tend to use the same walkways and food sources daily.

House mice generally breed year-round unless they live outside. Then, like the white-footed and deer mice, they breed in spring and fall. They can become pregnant early in life, before their third month birthday. (While that's quick, consider a house mouse's lifespan: one year.) While all the jokes about prolific breeders focus on rabbits, a female mouse can have as many as eight litters a year!

The males and females may stay together for a few days or for the whole breeding season, but generally they take new mates after winter passes. The female usually has a twenty-two-to-twenty-five-day gestation period before she gives birth in a nest she constructs to receive her young. The nest, a rough cylinder of shredded paper or other fibrous material about six inches in diameter, is usually stashed in a private crevice four to ten feet above ground, like a crack in a wall or a birdhouse—or it may be a hole in the ground. In a house, the nest could be in just about any sheltered location.

Baby mice, five or six of them, are born hairless and blind, but after two days they begin to sprout hair. After about a week, they have nearly all their hair, and they open their eyes after two weeks. Shortly after the second week, they begin to venture out of the nest, and they're weaned around the third week. They're sexually mature at six or seven weeks and set out on their own. The female will breed again after the young leave, and will continue to raise four or more families over the season.

Mice probably live about a year in the wild. Most predators will eat a mouse: fox, raptors, blue jays, snakes, coyotes, and bobcats, to name a few.

# VOLES

In the yard, mice are often confused with voles, who are vegetarians and eat their weight in plant food every day. Each female vole can have up to a hundred offspring in a year. Voles will girdle trees, eat bark around the entire circumference of tree and kill it. I don't know of voles ever entering houses, but they're sometimes a problem for farmers and homeowners.

Voles are stubby, stocky, brown or gray animals with dense fur and short legs and tail. They live outdoors in areas where they find lots of cover—grassy or littered fields or orchards.

Unlike mice, voles pose few health problems for humans, although they can carry some diseases: plague and tularemia. So think twice before you pick them up.

They eat grasses and forbs, seeds, tubers, bulbs, rhizomes, and occasional snails or insects. In the fall and winter, they're especially apt to eat bark.

Their home range is usually a quarter acre or less, varying depending on the population density, habitat, and food supply. They move around in a complex tunnel and surface runway system, often used by several adults and young. The runways are about an inch or two wide and are the most readily identifiable sign of voles. They're active day and night, eating and moving around.

Voles may breed year-round, but tend to do so mostly in the spring and summer. They have one to five litters a year and have three to six young each time. Young are weaned after twenty-one days, and females can breed at thirty-five to forty days old. They don't live for long, a price for a frenetic life-style, and an old vole is sixteen months.

You may first notice voles around your house when their population peaks, as it does every two or three years.

# RATS

Unlike mice, our society does not have any mixed feelings about rats. They're definitely low-life. Ever see a movie about a good rat? I'm talking about Norway rats, roof rats, and wood rats (the infamous pack rats). Well, the kangaroo rat is kind of cute (and is an endangered species). There's always an exception.

Norway rats are the ones we universally hate. When somebody says he saw a rat as big as a cat, he's probably exaggerating about a Norway rat he saw. They can reach up to a pound and have a coarse brownish fur and tails shorter than their bodies. And they're in every one of the lower forty-eight states.

Introduced accidentally around 1775, the Norway rat is also known as the brown, house, barn, sewer, gray, or wharf rat. Attracted to wharfs by the available food, it was no problem for the rats to access ships via ropes and then lounge about during the cross-Atlantic journey.

Not much good comes from a rat—unless you're studying behavior in a laboratory. They can burrow under buildings and damage the structure, gnaw electric wires or water pipes, gnaw through doors, window sills, walls, ceilings, and floors to get into a house. Norway rats carry leptospirosis, trichinosis, salmonellosis, and rat-bite fever. The plague, however, is more commonly associated with the roof rat, not the Norway rat.

Norway rats do well living around us in dumps, livestock buildings, silos, basements, sewers, and in our homes. They eat just about anything and plenty of it. Unlike human teenagers, they select a nutritionally balanced diet, fresh and wholesome, when they can get it. These rats need about an ounce of water a day, unless they're eating moist foods, like household garbage.

They're nocturnal, and like mice, they have poor eyesight, are color blind, and rely on their other senses for information. Their sense of taste is so keen that they often can recognize toxins in their food.

Rats eat about a half pound of food a day, and probably waste ten times that much. In an average day, rats travel an area about a hundred or a hundred fifty feet in diameter. They seldom travel any further than three hundred feet to get food or water.

A female rat bears four to six litters a year, successfully weaning about twenty offspring. They build nests below ground or at ground level and line them with shredded paper, cloth, or other fibrous material. Six to twelve young are born after a three-week gestation period. Although rats are born naked and helpless, they mature quickly, becoming independent in three or four weeks and breeding after about three months. Breeding peaks in spring and fall but lags off in the hot summer and often stops during cold winters, depending on where the rat lives.

Rats are pretty smart and adaptive, reasons they're so popular with behavioralist scientists. They'll avoid new objects placed in their environs so traps and poisons often don't work for a few days. They memorize all the characteristics of their neighborhoods, so people have to be pretty tricky to get a rat. They quickly learn what foods make them sick and avoid eating them again.

Not to be confused with pack rats, Norway rats often live in rat packs of sixty or more animals and are usually descendants of a single pair of animals. The packs may hunt together, and females will raise any orphaned rats in the group.

The roof rat, or black rat, lives on either coast and throughout the southeast. They're also known as ship rats and first arrived on this continent during the 1500s. They're somewhat smaller than Norway rats, weighing about half as much, with a naked tail longer than their bodies.

Their general biology is similar to the Norway rat's. Like other rats, they may horde or cache food, although they may never get around to eating it. They're more adventurous than the Norway rat and will often travel far in search of food, drink, or a home. They'll climb on utility lines, using their long tails for balance and will live in trees or high in buildings. Although the Norway rat can climb, it tends to live at ground level or lower.

Sometimes attempts to cull the Norway rat population will open up habitat for the roof rat. The more aggressive Norway rats will force roof rats out of the area or at least to different niches in the habitat. That's why you'll see infestations of Norway rats in lower levels of a building and roof rats in the upper.

The third pesky rat is the wood rat, a resident of the East coast, Great Plains, Northwest, and Southwest; they tend to live in rural areas. Their most notable behavior is a tendency to snag small objects—jewelry, pop tops, cooking and eating utensils, etc.

## Eliminating Rats and Mice

If you don't actually see a rat or mouse, you'll still know they're there. They leave droppings along their runways and in feeding areas—feces between a quarter of an inch long (mice) to nearly three-quarters of an inch (rats). Other telltale signs include oily smudges along their habitual pathways, and gnawed wood, food packages, windows, or insulation. You might be able to hear them gnawing or clawing inside the walls.

If you see lots of fresh droppings, it's a good indication of lots of fresh rodents. And they're not very careful in their bathroom habits. Also if you see rats during the day, it's a sign of an infestation, though seeing mice during daylight hours isn't a sign of infestation. A study at the University of Nebraska found that when sixty-five rats were found in an animal enclosure, only one rat was sighted every two and a half hours.

While you're thinking of exclusion techniques, consider some of these rodent Olympic feats:

- Rats and mice can:
  - run along or climb electrical wires, ropes, cables, vines, shrubs, and trees;
  - climb almost any rough vertical surface: wood, brick, concrete, screen, and weathered sheet metal;
  - crawl horizontally along pipes, conduit, or conveyors;
  - gnaw through a variety of materials, including lead and alum-

inum sheeting, wood, rubber, vinyl, and concrete block.

- Rats can:
  - climb the outside of vertical pipes and conduits up to three inches in diameter, and can shimmy up the outside of larger pipes by bracing themselves against the wall;
  - climb the inside of vertical pipes between one and a half and four inches wide;
  - jump thirty-six inches vertically and forty-eight inches horizontally;
  - gnaw and squeeze through an opening a half-inch in diameter;
  - drop up to fifty feet without being seriously injured;
  - burrow straight down into the ground for at least thirty-six inches;
  - reach at least thirteen inches along vertical walls;
  - swim about a half mile in open water, through traps in plumbing, and travel in sewer lines against a current.

- House mice can:
  - jump eighteen inches straight up;
  - travel hanging upside down from screen wire;
  - gnaw and squeeze through quarter-inch openings.

## Blocking Entrances

Before you do anything, find all the places the animals might use to enter your house.

Becca Schad, owner of Wildlife Matters, an integrated pest management firm in Virginia, says, "You really have to go over the house with a fine-toothed comb." Start by checking around the foundation, for cracks or openings, and block them with cement or masonry grout. Also unprotected openings like dryer vents are fair game to a rodent. If openings, even tiny ones, aren't closed, the rodents might be using them for an entry point, and from there, climb inside the walls to get access to virtually any room in the house. She says mice can get through cracks as small as a quarter-inch wide. Rats can gain entry through a half-inch opening.

Be sure to close off access around pipes and wires with cement, mortar, masonry, or metal collars. Metal baffles similar to the ones used to deter squirrels from getting to bird feeders will work along wires and pipes. Be sure to watch for signs of fraying and short circuits with any electrical wires though.

Another good mouse entrance is along the bottom of metal siding; mice will crawl up the unprotected ends. Use metal or mortar to block up the ends. They'll gnaw through rubber or vinyl weather stops.

Mice and rats may enter in a very mundane fashion, through the door. Doors should fit tightly with a distance between door and threshold less than a quarter-inch. It's easier to build up the threshold than to modify the door. If they're gnawing along the door, try installing flashing or a metal channel on the lower edge of wooden doors. Be sure it warps around the sides a bit too.

Rats are sneaky. Cover any floor drains with metal grates that are firmly attached. Make sure the openings are quarter-inch or less. Also, it's prudent to cover all pipes that leave your house, like the ones on your roof, with wire mesh. Many contractors fail to do this. You shouldn't. If people would take these preventative measure, "they would save themselves a lot of headaches," Schad says.

Often attics aren't airtight. Sometimes mice will climb up an outside wall, squeeze through a crack under the eaves, and set up housekeeping in the attic. Cover vents with quarter-inch hardware cloth.

Covering entrance holes is a necessary step, but it's a difficult one. Fill up all the cracks. To block up holes, first stuff the opening with steel or copper wool, then cover it with sheet metal. (This double line of defense reduces the likelihood that mice or rats will chew their way back into the house.)

I discovered just how persistent mice can be about getting into sealed places during a hiking trip in Idaho's White Cloud mountains not too many years ago. We left our cars in a remote spot—the middle of nowhere you could say. When we returned, one car had been invaded by mice, which consumed the bits of food that had been left in the car from the drive. The mice also tore apart paper products and fabric, stealing nesting material. It was easy to figure out *why* they got into the car. We never did figure *how* they got into the car.

Rats and mice have a tough time gnawing into a flat, hard surface, but they can bite into a rough surface with their paired incisors. That's why some materials just won't do for exclusion.

The U.S. Department of Agriculture recommends the following materials for rodent proofing:

- concrete, at least two inches thick if reinforced, three and three-quarters inches thick if not reinforced
- galvanized sheet metal, 24 gauge or heavier
- perforated sheet metal grills, 14 gauge
- hardware cloth, 19-gauge half-inch mesh to exclude rats, 24-gauge, quarter-inch mesh for mice
- brick, nearly four inches thick with joints filled with mortar
- aluminum, 22 gauge for frames and flashings, 20 gauge for kick plates, 18 gauge for guards

## Once Inside . . .

Before you make the house too rodent-proof, you might want to use some exit-only metal doors so the animals can have a chance to leave and not get back in. Just hinge the doors at the top so they swing out, but not in. They should be heavy enough to close firmly and fit closely enough so the rodents can't get them open again.

Once you prevent new animals from entering your house, you can concentrate on the home-dwellers.

Let's start with getting rid of mice. The method you use depends on how many mice you have inside. For the sake of sanity, I'll assume that you have a smallish family of about a half-dozen mice.

First, there's the old trusty method of a cat. It really works, and if you've ever considered getting a cat for any reason, this is one of the best.

Second, and slightly more aggressive, is to employ a Tokay gecko. The Tokay gecko is from Southeast Asia, and has a marvelous appetite for mice. You can buy them at most pet stores, and they're inexpensive. (As an added benefit, these geckos also relish large insects.)

Tokay geckos grow to a little more than a foot in length and have blue-green skin dotted with orange spots. They're nocturnal (as are the mice) and have no trouble walking on walls and ceilings, so even the most agile mouse won't escape them.

These geckos do have a few drawbacks, but in most pest-control endeavors, there's a tradeoff. Tokays walk around with their mouths open, and look rather fierce; they also bite when handled. Best to let them roam on their own. Unless you have tons of mice, you'll have to supplement their food. They're sometimes not very hardy in northern climates and need a warm, humid home.

Then, there are also traps. First of all, decide if your goal is to kill the animals or to capture and release them. If you want to capture and release them, make sure you release them far from your home, not across the street and not on the corner: They'll just come back to their home, the one you think is yours. Release them in a field where they'll find new quarters, far from those of humans.

The small Havahart mouse trap captures a single mouse, which can mean a lot of captures and releases. The rectangular box closes as the mouse enters to get the bait and remains closed until you release the mouse. A large-capacity mouse trap, like the Ketch-All or the Victor Tin Cat, doesn't have to be baited because the one-way treadles attract curious mice and then don't let them out of the trap. It probably would be better to bait the trap, though, to increase its effectiveness. The Tin Cat collects up to a dozen mice at a time. Check these traps often so the mice don't die of starvation or exposure. There are also live-catch traps for rats. Havahart makes one, as does National Tomahawk.

Glue board traps also work, although they're considered to work a little better for mice than for rats. Glue boards are like fly paper for rodents. When the traps are placed in runways, the animals get stuck to them and can't escape. Be sure to place them on beams and rafters to catch roof rats, and don't forget to treat the attic. Like you would for the live-catch traps, check the glue boards often. Kill the animals quickly and mercifully: drown them in a bucket of water holding them down with a stick, or use a quick blow to the base of the skull. If you have second thoughts about killing the animals after you've trapped one, you can loosen the glue with cooking oil. Glue boards are less effective in dusty areas or in extreme cold or heat. You can buy the traps ready-made or you can make your own with special glue and boards.

Now to the familiar: wooden snap traps. Nearly everyone knows these staples of cartoon slapstick humor. The theory is simple: The animal takes the bait, sets off the trigger, and gets squished by the metal bar that snaps down. In practice, it's not so simple. Rats, naturally fearful, may avoid the traps in their environment, and even mice are wary. To counter the rodent reluctance, put out baited but unset traps so the rats or mice can get used to taking the bait from the source.

Place traps abutting walls along the runways or travel routes with the trigger next to the wall. If you place double traps parallel to the wall, make sure the triggers face outward. For mice, set the traps close to their areas of activity, spaced no more than six feet apart. For rats, place traps close to the wall, behind objects, and in dark corners. If you enlarge the trigger with a square of cardboard, metal, or wire mesh placed just under the wire deadfall, the trap will be more effective.

Baits for rat traps include a small piece of hot dog, bacon, a nut, peanut butter, or a marshmallow. Mice baits could be a nut, chocolate candy, dried fruit, bacon, peanut butter, marshmallow, or even a small cotton ball, because mice are always on the lookout for nesting materials. For both mice and rats, the bait should be kept fresh and appealing because they won't be attracted to stale baits. Rats and mice will learn how to avoid these traps, so you need to be successful quickly: In other words, you need a successful first strike against mice and rats so set plenty of traps.

Traps are an effective—if labor-intensive—tool against rodents. Unlike skunks, you don't have to worry too much about getting bitten by a trapped mouse or rat; they're much easier to control. You can pick up a trapped mouse with a shovel or a piece of cardboard.

Once caught, dispose of the rodent quickly to prevent parasites from spreading. Never touch a dead mouse or rat with your bare hands, always use disposable gloves.

If you want to drive away the rodents, mothballs may work in an enclosed area. Just remember the fumes can bother humans too. If you know where the rodents are nesting, you can apply some of the sticky "Tanglefoot" glues used to discourage roosting birds and the rodents may be driven away. You may be able to frighten away the timid animals once or twice with a loud noise, but this method won't work for long.

Another strategy is to install an ultrasonic device. There are many on the market, and manufacturers of most of them claim the gadgets will drive the animals from your home. The devices get mixed reviews, but some experts claim they do work—under the right circumstances.

Lowell Robertson, president of Sonic Technology, which manufactures the Pest Chaser ultrasonic devices, says, "Ninety percent of the ultrasonic devices sold are ours." He says it wasn't until the late 1970s that chip technology developed to the point where it was possible to make high-quality ultrasonic devices at low prices.

He says rodents are a primary target for ultrasonic devices. "All the rodent family, common rats found in the United States and really all over the world, have the ability to perceive ultrasound," he says.

Ultrasound devices are based on the fact that different creatures hear different sound frequencies. Sound covers an extremely broad spectrum, going all the way from subsonic sound to the auditory range to ultrasonic to microwave frequencies. Human beings can hear sounds of up to about twenty thousand cycles a second; that's as fast as our eardrums will vibrate.

Rodents can hear sounds of up to 100,000 cycles per second. However, within that range there are two different specific frequency bands that rodents hear best. One is just above the human range at 21 to 23 kilohertz. However, people complain about being able to sense the sound at those frequencies; those sounds give people a headache. It's even worse for dogs and cats; those frequencies "would blow peoples' dogs and cats out of the room," Robertson says. As a result, those frequencies aren't used for ultrasound devices.

The other range that rodents hear well is right around 46.5 kilohertz. Robertson says his device broadcasts in a range from 32 to 62 kilohertz, 1,000 cycles per second. This is an optimum range to use "because we can spike it so it hits right on what rodents hear very well, yet it's beyond what dogs and cats can hear, and above the ranges that trigger TV sets, garage door openers, and ultrasonic burglar devices," he says.

"All we do is create acoustic stress," he says. Robertson explains: "You know the big speakers that are right in front of the stage at a rock concert? Can you stand being in front of one for very long? That's 110 dB by law; that's the maximum that they can put out. The decibel scale is like the Richter scale, it's logarithmic. A 10 dB increase is a ten-fold increase in sound pressure level. So a 120 dB is ten times louder than 110 dB and 130 dB is a hundred times louder than 110 dB. And 140 dB is a thousand times louder. Our device puts out 140 dB. Could you stand being in a room with one of those speakers that was one thousand times louder than that rock concert speaker?"

Neither can the rodents, he says. "We create an acoustically hostile environment: We're changing the environment by adding an aspect that those rodents cannot and will not deal with."

To make things more unpleasant for the rodents (and to make the devices more effective), they no longer just generate a single tone, he says. Many of the devices that were created early on and failed in the marketplace are what is known as single-tone generators, he explains. "They would broadcast say, 50 kilohertz real loud. After a time, rodents will get used to the sound; they'll live with it. And after a couple of generations, rodents will produce offspring that are deaf to that tone range."

However, Robertson's devices now create a complex sound, one "that's going through 32 to 62 kilohertz in a wide frequency band, and we pulse that sound at sixty times a second in a sinusoidal wave pattern, which is a nonrepetitive pattern so it's changing all of the time." Rodents don't get used to that, he says. "That's what it takes to make a good ultrasonic device: loud noise, complex sound, constantly on," he asserts.

Robertson also says consumers need to put a device in each room where there is a problem, because the sound will not pass through any hard surfaces. "In fact a sheet of paper will stop ultrasound. Ultrasound is very fragile, very directional. It goes in a straight line from the broadcast point. If it hits a hard surface, it will bounce back, and you'll get a carom effect, like on a billiard table. So you can put one high-output device in a pretty good sized room three to five hundred square feet, and the bounce around will fill in the whole room and it will give you protection in that room."

However, he points out that the sound won't go beyond that room. "That's one of the major mistakes in claims that a lot of ultrasound companies made, and you'll see it in some of the catalogs. They say it covers thirty-five hundred or twenty-five hundred square feet.

That leads the consumer to believe that one device will cover a twenty-two-hundred-square-foot house. Basically you need a device in each room where you have a problem potential—where food is prepared, consumed, stored—kitchens, family rooms, basements, garages."

Another method is poison, and at one time or another you might be tempted to use poison. Don't. There's always the possibility that some unintended animal will eat the poison. And although it's rare, children are still killed by poisons left for rodents. More likely, however, is the probability that the mice or rats will eat the poison and die in your walls. Where you will never get to them—but you'll know that they're there. First you'll hear frantic scratching from within. Then you'll smell them.

Poisons are difficult to use, too. Rats may not eat them or just eat enough to get sick and learn to avoid the poison. And rodents, like insects, gradually become immune to the chemicals designed to kill them. Rodenticides really aren't the way to go for the household pest chaser.

Whatever method you use to rid your house of rodents, you can then help keep them away by reducing their food sources. This means regular vacuuming, mopping, sweeping. This means sealing food in jars and other airtight containers. Metal and glass are the only materials that count. Cereal in their original boxes do not count. Rodents will smell the cereal no matter how tightly you curl the wax paper. Hungry mice—or rats—will gnaw through the box and then use the cardboard for nesting material. Don't leave pet food out for longer than you must—rodents consider themselves pets, too.

## Attacking the Yard

If your yard provides a good habitat for mice, it's only natural for them to explore your house. Discourage those tendencies by making the yard abutting the home inhospitable to mice: Keep grass mowed, engage a seed catcher beneath your bird feeder to prevent seed from reaching the ground, and clean up pet food. Raise your woodpile by fifteen inches; mice won't burrow under it then. Keep field mice in the field.

Clean up any cover and litter. High grass and heavy brush attracts a variety of animals. Keep bushes and trees trimmed so they don't touch the house. If you can, cut brush three feet back from the side of the house. Heavy vines and vine ground cover are also good cover for rats. Rats also need more water than house mice, so get rid of standing water.

Metal trash cans are better than vinyl or plastic ones, which the rodents can gnaw through, and it's best to have the cans in a stand so they don't tip, releasing attractive garbage. Of course, don't feed your pets outside. In the house, stop leaving uneaten food in the pet's bowl. The dog or cat may complain for a while, but it's best to do away with the attractant at least until the rodent problem is under control.

## Eliminating Voles

For a homeowner, dealing with voles should be fairly easy. When voles eat bark, they leave uneven gnaw marks—at varying angles and in irregular patches. By contrast, rabbit gnaw marks aren't distinct like the vole's—about one-eighth-by-three-eighths inch. If they're bothering a few trees or seedlings, you can protect them with quarter-inch hardware cloth cylinders buried about six inches into the soil.

You can generally get rid of voles by altering the habitat: Eliminate weeds, ground cover, and litter that provide food and cover. Mow the lawn regularly and keep mulch about a yard away from tree trunks so they can't get close under the cover and eat. Tilling the soil will destroy the runway and tunnel system, and probably reduce the population somewhat.

### ❖ ROUTING RODENTS

1. Block entrances with rodent-proof material.
2. Dry up water sources.
3. Store food in rodent-proof containers.
4. Cut down brush around the house.
5. Keep available outdoor food—garbage or pet food—locked up tight.

# PIGEONS AND OTHER BIRDS

When I was a kid, I loved the scene in *Mary Poppins* where the old woman sings "Feed the Birds" about sweet little birdies who sit on your finger and sing, looking up at you adoringly. They were always the good guys in Disney films. There were lots of birds in my life. Tweety Bird. Big Bird. Woodstock. I liked them all.

Then one day when I was visiting Busch Gardens in Virginia, a seagull swiped my pizza. Then I noticed all the pigeons perched on the statues, defacing them with droppings. And some starlings built a nest over my front door, dropping straw, grass, and other debris on my head daily. My world ceased to be black and white, good and bad. Birds became both lovable and loathsome.

While always welcome, birds are sometimes inconvenient. It's not that I mind a bird's nest—I just don't want it over the front door. Other people have different levels of tolerance.

At this juncture I want to point out that I'm not only a bird feeder, but I am also the author of two books on how to get more enjoyment out of feeding wild birds. Most birds I like. Some I don't.

# PIGEONS

Almost everyone has an opinion about pigeons.

But no matter what your opinion of pigeons you must marvel at them. In the harshest urban environments, pigeons are prolific. Pigeons are a part of Venice's charm. The Swiss Army is giving pigeons new skills by training them as two-way messengers. Even the most ardent pigeon-hater is soothed by their lyrical cooing. Pigeons are smart. Pigeons are feisty, trainable, and persistent. But they are nice birds, cordial to each other, and even tolerant of cats.

Pigeons, their technical name is *rock dove*, frequently make the first page of newspapers. Here's a headline from the first page of the *Los Angeles Times* metro section, November 26, 1989: PIGEON TRAPPING AT CONDOMINIUM IN LAGUNA BEACH RUFFLES SOME FEATHERS.

Pigeons are not indigenous to the United States, but probably were carried by European settlers for food during their long voyages at sea. Because a boat would occasionally arrive ahead of schedule, a handful of pigeons might escape the dinner plate. Food being abundant in the New World, these pigeons, which arrived in the 1600s, were allowed to fend for themselves in the wild. Almost immediately pigeons took up residence around human settlements because these sites provided an abundance of food scraps for the birds, and because pigeons survive only in open areas.

Pigeons can eat almost anything—grain, bread, crumbs, seed. They thrive on handouts and garbage. Handouts, offered by "nice" people and tourists amount to as much as seventy-five percent of a pigeon's diet. I was shocked, but not surprised, to see a middle-aged man one summer afternoon sprinkle an entire bag of birdseed on the corner of a neighborhood parking lot—a favorite congregating place for pigeons. This community of pigeons, which I estimate to be about two hundred large, flourishes because it is well-nourished by well-meaning, but kooky, people.

Almost three hundred years since its introduction, the American pigeon is rarely regarded as an important source of food. Yet the bird is controversial. When the Laguna Sea Cliffs Apartments in Laguna Beach, California, decided to rid itself of a flock of about sixty birds that were napping on people's balconies, the apartment owners created an uproar. The condo owners wanted to trap the birds and send them to pigeon heaven. But before the program could start, somebody discovered that it was against city law to harm pigeons in any way. Besides, many people in Laguna Beach liked pigeons. A prominent citizen said, "We have to be more tolerant of nature and redefine what is a menace."

How to deal with the problem of soiled balconies? One suggestion, giving pigeons birth control pills, was quickly discarded because squirrel lovers would revolt (the pills would affect squirrels, too). In addition, as one activist put it, "If you get into birth control, we're going to have the pro-choicers out here." Relocating the pigeons was considered and tossed out, too, when people realized that the pigeons would "beat you driving back."

Pigeon racing, according to the *Los Angeles Times*, "has been a little-noticed part of city living for decades." The ancient sport, however, isn't popular with all city dwellers: In Torrance, California, the community group RAP—Residents Against Pigeons—was founded to put a stop to the sport. (The Torrance city council refused to outlaw pigeon racing.) Roger Mortvedt, a racer, calls these pigeons "highly bred athletes, born for racing. They don't hang out on fences and wires messing up the neighborhood, because they're trained not to do that. You can win or lose a race by seconds, so you don't want them to get in the habit of hanging around." Pigeon races are several hundred miles long and can involve fifteen thousand birds. While not thoroughbred horses, the fastest pigeons—those breaking sixty miles per hour—can command a price of one thousand dollars.

Physically, pigeons are remarkable. They are intelligent, strong and swift. Many of their senses are superior to humans': pigeons can hear ultrahigh frequencies, see ultraviolet light, and "sense" magnetic fields. They can sustain a sixty-mile per hour flight for an entire day. And, as you may have surmised, they are fertile: At six months, female pigeons can have six to eight baby pigeons—every year for the next seventeen years. Pigeons are able to withstand great stress. Witness what happened to this British pigeon:

A greedy pelican, apparently with a penchant for smaller feathered creatures, snatched an unsuspecting pigeon in St. James's Park, London, yesterday and tried to gobble it up. At first, things looked bad for the pigeon. It disappeared inside the pelican's pouch at one point. But a savior was in sight. A small boy among the crowd of onlookers distracted the pelican by swinging his arms and the pigeon managed to flutter free into the sunshine. [*The Daily Telegraph*, May 20, 1989]

The image of the pigeon feeder being a homeless old lady isn't true. Pigeon feeders, a large subset of the eighty-million-plus bird feeders around the country, are passionate about their hobby. Sometimes their passion stems from having to be so defensive about feeding pigeons, sometimes it comes from love for these gentle birds. Feeding pigeons grain and seed most closely approximates their natural diet, but people insist on feeding them bread, gruel, and table scraps. If pushed to it, pigeons will eat garbage, insects, and even livestock manure.

In the wild, that is to say in the middle of a city, pigeons breed any time of year. Whenever they feel like it. Even in winter. Pigeons typically produce two eggs and nest several times a year. Nests get built just about anywhere flat enough for roosting, and can take the form of crude platforms made of twigs and grass.

Both males and females care for the young, who hatch out from eggs about eighteen days after laying. The squabs feed on pigeon milk, secreted by the parents, and are weaned after four weeks or so. Never ones to waste time, the parents may lay new eggs before the nest is empty. Pigeons live about three or four years in the city. Funny, it seems like they never die, they only increase their population.

Well, you know what they say about beauty. At the 1989 Illinois State Fair 630 pigeons were entered in—are you ready for this?—a beauty contest! Thirty different kinds of pigeons entered, some of which looked like "miniature white peacocks . . . and others [so] fat [they] could eventually be called 'squab' on a restaurant menu," said the *Chicago Tribune*.

The truth is that most people don't like to have pigeons around. (However, would anybody like our cities to be completely devoid of nonhuman animal life?) Pigeon droppings exude an acid that eats into building stone. Large amounts of the droppings kill vegetation, not to mention your nasal passages. The droppings are ugly and pigeons seem to have uncanny aim when you are wearing expensive clothing. The birds themselves carry a host of unpronounceable diseases that include tuberculosis, salmonellosis, histoplasmosis, toxoplasmosis, psittacosis, cryptococcosis, meningitis, and encephalitis. Mites, fleas, and ticks on pigeons will also nibble people and can be pests in the home.

There are plenty of reasons for not wanting them roosting around your home, but the pigeon is a steadfast, if not ferocious, foe. In their quests to coax pigeons into roosting elsewhere, people have tried all sorts of concoctions. Noisemakers, Nixalite (also called porcupine wire), low-voltage electric wire, ultrasound (which unfortunately has been known to affect heart pacemakers), and mothballs are often used.

But one of the most common techniques is to mount a plastic owl, because pigeons are deathly afraid of owls. These owls are pretty popular; so too is the sight of a pigeon perched on top of the plastic owl. The trick is to keep the owls in motion. Experts agree that the plastic predators must be moved often, but their definitions of often vary: Some recommend moving the replicas every few days, and others recommend moving them every few weeks. While pigeons are not known for their intelligence, they won't take long to figure out that the stationary owl is no threat. And once the birds realize the owls can't move, let alone eat, no amount of movement is going to make them change their minds. Cal Saulnier of Plow and Hearth catalog says his company has sold some Great Horned Owls to the Department of Defense. Rumor has it that birds were making a mess of Blackbird secret spy planes and the owls were drafted into service to keep the birds out of the plane hangers. How did they work? Major Cole with the U.S. Air Force will call me and let me know.

Some cities are taking a more ecological approach to pigeons: To reduce the herd, peregrine falcons have been reintroduced in cities. The falcons are thriving. But so are pigeons. In reality, falcons probably eat one pigeon (or rat) every week, and the territorial falcons need a lot more space than do either pigeons or rats, and falcons have a slower reproductive rate. Falcons may not drastically reduce the pigeon population, but they do help.

Nixalite and noise are probably the best two offensive weapons against pigeons. Nixalite is like a bunch of sharp jacks strung together, giant barbs pointing in all directions, that prevents pigeons from alighting. Pigeons are easily startled, though they quickly return to their original place. Continual noise through radios—talk radio stations are the best—or wind chimes (assuming you live in a windy place) are good tools. Firecrackers can't be beat, but you usually have to keep tossing them. If you're lucky an occasional loud firecracker will scare pigeons away indefinitely. You have to keep up harassment until the pigeons find new roosts. Since pigeons are usually more of an urban problem, devices that make continuous noise or intermittent explosions may do more harm to neighborly relations than good in pigeon frightening. Gas exploders, alarms, and shotguns work, but not in the city.

Exclusion takes a little more work. Pigeons roosting in indoor areas need to be pushed out: Try mothballs or flakes or loud noises, and then block vents, eaves, and other openings with quarter-inch mesh screen, wood, glass, or even plastic netting. Mothballs can be toxic to humans, so be careful using them around people. Just make sure you're not trapping any birds inside when you close the exit.

Netting is effective not only against pigeons, but against sparrows and starlings as well. Just about any structure you want to keep pigeons off—buildings, ledges, detached houses, cars, even trees—can be covered with netting. When you see netting under bridges it usually isn't there to prevent workers from falling, but to keep pigeons from roosting, a noble cause.

There are other techniques you can employ to prevent pigeons from hanging out. A single wire "fence" several inches high may help prevent pigeons from roosting on a roof or other surface. Electrifying the wire will encourage the birds to go elsewhere. A low-voltage, low-amperage current that won't harm the birds will be sufficient. Pigeons will perch on flat surfaces, but not ones at more than a sixty degree angle. Use wire mesh, metal, or wood to create angled surfaces. In open buildings with exposed beams, use netting or mesh to block access to the upper reaches of the building.

Nontoxic repellents make roosting uncomfortable for pigeons. The tacky repellents come in a variety of forms: dry, spray, paste, and liquid. But you have to use these repellents on all surfaces in the area, or the pigeons will just bunch up in a corner away from the sticky area. When you apply the repellents, make sure you don't leave any more than three inches between any strips of the substance, or the pigeons will stand there! If it's not too dusty, an application should last about a year. It may stain some surfaces, so place it on painter's tape or something you can later remove.

If you need to destroy pigeon nests that are left behind, be careful. Laurie Bingaman, on staff with the National Zoo, reminds people to take precautions against airborne diseases carried by pigeons: "Get a towel, get it damp, and tie it around your face like a bandit would, covering your mouth and nose. Then the area the pigeons have been living on, any nesting material, and any droppings should be dampened gently to avoid stirring up dust. Dispose of the mess in a plastic bag. Finally, clean the former pigeon area with a disinfectant." With the high reproductive rate of pigeons, it does little good to destroy pigeon eggs—it's an insignificant action, destroying one or two eggs when they're so easily replaced.

John Adcock, owner of a pest control company in Maryland, is a little more cautious: "The real problems are pigeons. They're filthy animals and leave bird droppings all over. You've got the histoplasmosis. There's about twelve different lung diseases you can pick up from these critters. We use respirators—the same type of respirators you'd use if you were cleaning up hazardous waste such as asbestos."

"If you can eliminate the habitat that attracts the nuisance animals, you've corrected the problem and you've corrected it in an enduring and fundamental way. There's a lot of solutions that are very short-term; a lot of exterminating companies are big on short-term solutions because it creates a constant business for them. The fundamental solution is usually habitat alteration if that's possible," recommends Al Geis, research director of the Wild Bird Centers of America.

Geis once worked on a bird project for the state of Pennsylvania to help eliminate a pigeon problem on a bridge over the Delaware River to New Jersey. "I ended up walking the catwalks underneath these things, and I noticed these piles of pigeon remains; there would be pigeon intestines and occasionally pigeon heads, and I asked the work-man 'What's that all about?' and he said, 'The hawks leave that be-hind.' Well, they had a peregrine population living in these bridges, feeding on the pigeons."

A pest control operator proposed baiting the pigeons with corn on the New Jersey side (Pennsylvania had outlawed the practice) and then substituting strychnine-treated corn for the straight corn to poison the pigeons. "My principle contribution to this project was to point out it was contrary to the provisions of the Endangered Species Act to even place those birds [peregrine falcons] in jeopardy," says Geis.

While the predators had no significant effect on the pigeon population, Geis noted that "Poisoning the pigeons wouldn't significantly control the pigeon population as long those nice, broad flat surfaces in a protected situation remained." He recommended solutions that would exclude the pigeons from the bridge. For instance, pigeons can't perch on a thin wire, so if you string a thin wire about an inch or an inch and a half above the surface where the pigeons are roosting, they'll find another spot—away from you. Just make sure the wire is taut and doesn't touch the surface it is protecting.

"These animals have tremendous reproductive potential and as long as the area is attractive to them and provides suitable habitat, they are going to use it," explains Geis. Which is to say that under normal circumstances you won't be able to kill enough of the birds to significantly control the problem, but you can exclude them, or largely exclude them, from particular areas.

Geis suggests that a big step in outwitting critters may be changing the way humans view the problem. For example, if a sharp-shinned hawk is attacking and eating the songbirds at your feeder, perhaps the best thing to do is to let nature run its course. "It's not a real problem, it's just a wonderful opportunity to see sharp-shinned hawks. And the songbirds have great reproductive rate and high mortality to balance. If that sharp-shin doesn't eat a bird around your feeding situation, he'll get one somewhere else."

If you see someone feeding pigeons in your neighborhood . . . well, good luck.

## STARLINGS

They're pigs at the bird feeders. They roost in huge numbers, fouling ledges and sidewalks. Worst of all, they get to your strawberries before you can.

Like their fellow urban habituate, the pigeon, starlings are an introduced species, first coming to our shores in 1890 and 1891 by a man who wanted all birds mentioned in Shakespeare's works to have a home in America. He didn't get them all here, but the starling is his great success. By the 1930s the birds were spotted in Nebraska, and then in Colorado nine years later. They're so successful here that they're now found across the United States, in the southern half of Canada, in northern Mexico, South Africa, Australia, and New Zealand. Why do starlings do so well? For one thing, they thrive in open areas like cities, towns, suburban developments, and farmland.

A big reason for their success is that they eat almost anything— seeds, fruits, insects, and garbage. Probably half of their diet comes from insects. Like pigeons, they are successful at reproducing themselves. They make nests just about anywhere, in buildings, tree cavities, outtake pipes, or birdhouses intended for other tenants. They nest twice a year, laying four to seven eggs. Young starlings leave the nest after three weeks.

Like crows, they roost together in huge flocks when not nesting. You're most likely to see them gathered together during the winter in dense cover out of the wind. Even though they leave the communal roost during the day, they return at night, sometimes flying thirty to fifty miles round-trip. Some starlings migrate each year, while others stay put.

Starlings are about the size of a robin, and they're dark brown with light speckles all over. The biggest objection people have to the birds is that they aggressively push out native species, most notably the bluebird, flicker, purple martins, and other birds that nest in cavities. They push other birds away from feeders, and then invite their starling friends to dinner. They're messy eaters and waste seed on the ground.

The best weapons against unwanted starling roosting are Nixalite or Cat Claw and a sloped ledge (both of which are outlined in the section on pigeons). Wire mesh or netting installed on the underside of rafters keeps them from roosting or nesting in the upper reaches of outer buildings. Starlings find tacky repellents objectionable, so the products will keep the animals from roosting and nesting in an area. Try startling starlings. Use gas cartridge exploders, mylar tape, fright eyes (similar to the scareeye), recorded distress calls, bright lights, plastic owls, firecrackers, and drums in combination. Vary the intensity, frequency, and location of your attacks. One thing we have in common with starlings and other birds—neither of us can hear ultrasound. If these methods don't work against starlings in the garden, try bird netting.

Starlings are a little more aggressive than other birds in trying to gain access to our houses. Becca Schad, owner of Wildlife Matters, a Virginia humane animal control company says, "Starlings will often poke a hole in window screens with their beaks, make a hole, and then the squirrels follow." Plain window screen won't do; it's not strong enough. If starlings are trying to build a nest, just clear everything out and cover the opening with hardware cloth. While Schad won't kill baby starlings, "It can't help it if it was born in your attic. Eggs, I don't feel too bad about destroying, you have to draw the line somewhere." And it's best to stop things early, because starlings reuse the same nest.

Making sure there are screens on the attic vents and the other vents in the house. Sure it's a hassle, but consider the alternative. Schad often gets springtime calls from homeowners who want her to remove young starlings from stove vents when the young ones get trapped in the duct. "When the babies get big enough to fly, instead of going outside, they fall back in the duct, and then I have to take the fan out of the stove and take the babies out that way and take them out through the kitchen."

## Pigeons and Starlings at Wolftrap Park

The staff at Wolftrap Park, a national park dedicated to the performance arts, have been locked in a fierce battle of wills with birds for years. In 1982, when the original theater burned, the Park Service had a chance to amend the situation by redesigning a new theater. Chief Ranger Bill Crockett says, "In that building there were many, many nooks and crannies and ledges, and birds loved the environment. So when they built the new theater that was one of the new things that everyone asked for, that it be bird-proofed as well as possible, and the architect did try to design a building that would provide fewer little nooks and crannies. However, these birds are so inventive that even on the narrowest little thing, they can build a nest. So what has happened is that we still have a nest problem. It's not to the extent we had in the old theater, but still it's a concern and something we have been working on ever since the theater went into operation in 1985."

While they have been able to deflect the pigeons to some extent, Park Service staff haven't come up with a way to get rid of the starlings. Starlings seem to nest about anywhere. At Wolftrap they have found nesting habitat in take-up reel boxes, which were designed specifically to keep birds out. The reels raise and lower the speakers over the front orchestra, and the fronts of the boxes where the cables come out are protected by a set of stiff brushes. Yet the starlings squeeze in between these brushes and nest inside there. "I mean it's almost impossible!" says Crockett.

Unlike pigeons, starlings can land on just about any surface. "Starlings go just about anywhere. They are not our major problem, because their droppings are smaller. Pigeons are our B-52s." When you speak with Crockett, you hear a lot of military analogies.

"It's like a losing battle. It's an outdoor theater and as much as a natural area as we can make it, but people don't understand when they spend twenty-five dollars for a front orchestra ticket and they get a bird dropping on their three-hundred-dollar dress.

"We've put up a product called Nixalite, which is a spiny sharp wire that birds don't like to land on because it pricks their feet." Each year Park Service staff climb into the nether reaches of the theater to put up Nixalite—or clear it off, because birds drop nesting materials on top of the Nixalite and build it up so they can sit on it. Staff also stuff any holes with steel wool, or block off roosting spots with hardware wire.

"Birds won't stand on the piano wire stuff. We use it over some railings. The Nixalite is actually easier to put up and maintain, I think, than that piano wire. Certainly the Nixalite is less attractive, but its easier to rig up the Nixalite than the piano wire on some of the stuff we have to protect." Sometimes only repetitive manual labor will suffice. House lights that shine upward are protected around the edges with Nixalite, but the birds will still nest on top of the bulb. "We can't Nixalite the whole middle. So in the spring we have to go up there and remove the nest because when the lights go on, the nests catch on fire," says Crockett.

"The other thing we've done is to close up little holes wherever we can by filling them with steel wool or covering them entirely with a sheet of metal. The upper part of the cover spot booth, which is the booth that hangs over the balcony and contains spotlights, is completely covered with wire mesh to keep the birds from roosting on the flat surface of the booth top.

"We feel that we've eliminated about eighty percent of the problem just through these means," says Crockett, but he goes on to say that he is beginning to consider draping a net over the entire theater while it is closed from fall to spring. The net would discourage and virtually stop most birds from getting in the theater, particularly in the early springtime when they're looking at their first nesting period. Ordinarily birds that have two nesting periods will return to their first nesting place for their second nesting. "Once activity starts up in the theater too, around mid-May, and certainly by the end of May when performances start, you have less bird activity because of the noise and all the disruption. But birds that get in there in the early spring period and start building a nest aren't going to give up their homes, not for the loudest pop rock or whatever; they're going to settle in and get used to it. In 1990 we covered the speakers with wire mesh and have seen a drop in bird population."

The Park Service has tried many strategies at Wolftrap. "We put up plastic owls; we used plastic snakes. We found they did not work. If they don't move, the birds know it. They get used to it and they realize it and know it's not real," says Crockett. "The birds were sitting on the plastic owls' heads."

In spite of everything, Crockett retains a positive outlook. "There's room to support a certain number of pigeons. The thinking is that if we could go in there now and kill every bird that's in there now, that the population would be down a year or two and then start going up again. The best way to eliminate them is to keep them out."

## SEAGULLS

Although you visualize beaches when you think of seagulls, they're dispersed everywhere, from coast to coast and even inland. Salt Lake City even has a monument to seagulls, the animals that saved residents from a plague of crickets, arriving at the last minute and devouring the pests. In other cities, gulls are less welcome as they dive into garbage dumpsters and plague customers at outdoor eateries.

There are seventeen species of North American gulls, but the most common include the laughing gull, Franklin's gull, the great black-backed gull, the California gull, and the ring gull, found around the Great Lakes. They've adapted nicely to human presence, and many species are actually increasing their numbers: The ring-billed gulls of the Great Lakes have been increasing their population about ten percent a year since the early 1970s. Since they're classified as migratory birds, people can't hunt, trap, or kill them without a special permit.

Seagulls are omnivorous and eat carrion, mollusks, small mammals, food from humans, and garbage. Herring gulls will stuff themselves until they can no longer walk, let alone fly. It's not that they're gluttons, it's an adaptive behavior from times when they went through periods of lengthy famine (an unlikely occurrence today with the availability of food in dumps and fast food parking lots!).

Seagulls are such successful scavengers that they've earned the title "flying rats" in many parts of the country. No longer content with dumps and garbage cans, they're so bold as to pirate away snacks and meals right out of our hands. Many beachfront resorts in Florida protect their pool and café areas with an enormous string cat's cradle rigged above the areas. The string zigzags wildly among the palms and poles to form a barely discernible maze. The Don Caesar resort in St. Petersburg Beach, Florida, found that the strategy reduced seagull incidents by fifty percent.

Dave Sileck, beach activity director for the Don, explains, "The only secret is to use it in a very random fashion so there is no pattern. Geometrically, it will work out to be as many angles as possible. Seagulls are pretty smart; they do eventually figure it out, so you either have to add some line or take some line away every once in a while to keep them honest. They don't see the wire, they just feel it. We use a number 10 gauge fishing line, which is very thin. They feel it on their wings and they get freaked out and they just take off. But of course if there's enough food there, they'll do anything for that." Of course, they can land outside the cat's cradle and walk under it, too.

Commercial fish ponds have long used a similar arrangement, employing .4 mm steel wire strung in parallel spans eighty feet apart over the ponds. Dumps have used it, but since the attractant, rotten garbage is so alluring, they had to place the wires closer together, at fifteen feet. Fast food restaurants use even closer parallel spans of monofilament wire. Although the strategy does seem to work for gulls, no one knows why; and the strands won't exclude pigeons or other birds. Seagulls will also avoid porcupine wire and sticky bird repellents.

If cosmetic concerns about birds seem frivolous, consider that gulls are more likely to collide with aircraft than any other bird. Worldwide, there are some ten thousand bird strikes every year, mostly over wetlands, prime bird habitat, and 140 people have been killed from bird-plane collisions over the years. One million ducks, five million geese, a half billion blackbirds and starlings, and 700,000 gulls fly the friendly skies along with the planes over eastern North America alone.

The first gull-plane crash occurred in 1912 when transcontinental pilot Cal Rodgers and his plane collided with a gull. Both were immediately killed. In 1978, a DC-10 taking off from Kennedy Airport hit a flock of seagulls. Although the plane suffered severe damage, it was able to land and crew and passengers were evacuated before it exploded into flames. In 1977 a private plane taking off from Chicago's Meigs Field hit a flock of seagulls. The pilot and three passengers were immediately killed, and the airport crew cleaned up 180 dead gulls.

Aside from being three scary stories about gulls and planes, these examples point out that airports must have strategies to deal with birds. Airports have to disperse the birds who want to loiter on runways, and they use many of the same tools you can. They use gas exploding devices, recordings of distress calls, flares, and exploding or whistling shells. Officials take care to eliminate food, shelter, and water. (Kennedy Airport drained decorative pools that attracted the birds.) And they keep up scaring techniques at night. An airport in

New Zealand uses model airplanes in the form of hawks to scare off birds. The Royal Navy drafted falcons to keep runways clear. Some airports mounted dead or model gulls in contorted positions along runways. What they found was that the techniques worked best when combined with others and used diligently. Check the resources section at the back of the book for suppliers of various devices for discouraging birds.

# HOUSE SPARROWS

Another introduced species, the house, or English, sparrow, makes itself unwelcome at bird feeders and birdhouses when it bullies the other birds. House sparrows were released in 1850 in New York and have since moved from shore to shore in the United States and throughout southern Canada. Well, they're here to stay now, and we need to figure out ways to live together.

I have trouble telling house sparrows from our own native sparrows, but if you can get a good look, you'll see the black bib on the male. He's the easiest to spot. He has white cheeks, a gray crown, and a tan neck. The female is kind of washed-out. She is a streaky dull brown on her back with a dirty white underbelly.

House sparrows eat mostly plant material, but about four percent of their diet comes from animal food. In areas around people, they supplement their diet with highly processed foods from garbage, bread crumbs on lawns, and refuse from outdoor eateries. Baby sparrows thrive on a high-protein diet; about sixty-eight percent of their diet is insects and the rest comes from plant material.

The birds can breed throughout the year, but are most active between March and August. A sparrow pair builds the nest and rears the young. The nest is a sloppily constructed version of the other nests in the weaver family, to which the sparrow belongs, and it has a roof. The male selects the site and protects the territory around it, and the female incubates the four or five eggs. They hatch in about two weeks, and the young are ready to fly after about another two weeks, although the parents may continue to feed the fledglings even after they're out of the nest.

House sparrows don't travel far from their nests during the nesting period, but nonbreeding adults will travel miles to a seasonal feeding area. They live about five years in the wild, and the first year is the most dangerous, with the highest mortality rate.

The greatest problem with house sparrows is that they push out other native species. There's not much you can do to keep a house sparrow away from your feeder, because feeders designed to exclude large birds like starlings will allow finch-sized birds like the sparrow to feed. In theory, they're more fond of a mixed feed than just sunflower seed, and they won't eat thistle at all.

Keeping them from birdhouses is a little tougher. They love martin houses, and will move in before the first martins return from winter feeding grounds. Try blocking up the holes in the early spring so the house sparrows can't get in first. Your martin house should be high on a pole, away from obstructions so the martins can have space for their acrobatics. Also, you need to clean the house each season after the martins leave, or they won't come back. House sparrows can't get in openings of one and one eighth inches or less, but these houses can be used by wrens, and you can buy or make special bluebird houses. Often house sparrows don't like bluebird houses because the floor size is small. Other suggestions for a bluebird house include making the door one and one half inches in diameter and putting a three-and-a-half-inch screened hole in the roof. Apparently bluebirds don't mind a summer rain, but sparrows do. Watch birdhouses carefully to make sure sparrows don't move in. If they do, tear out the nests, daily if necessary, until they give up. Also place the birdhouses away from human dwellings where the sparrows like to congregate—bringing us to another problem with sparrows.

You might find you have too much of a good thing with sparrows hanging around the house, roosting and building nests. You can exclude them with netting or porcupine wire or you can try a sticky repellent.

Niles Kinerk, director of the Gardens Alive! Gardening Research Center thinks sticky repellents work. He says, "I'll tell you what we have used, and it's extremely effective depending on what you're trying to do, and that's Tanglefoot. It's amazing. If you've got roosting sparrows that you don't want, and they're making a mess, put down a string of Tanglefoot, and it's over. It works on any bird; if they get their feet in this, they don't come back.

"We had an overhang over our entrance, and it had a corrugated metal cover with all these wonderful little spots where sparrows could nest, and of course they made a mess where everybody was coming into the building. So we put a string of Tanglefoot around the spot and had vengeance." The birds evidently found another place to nest, because they abandoned the spot over the door.

# HERONS

I spent a week on an island in the Great Barrier Reef off the Coast of Australia. The island was called Heron Island. At that place, home to more birds—including herons, of course—than I have ever seen in one spot before, you had to wear a hat all day long. And it wasn't because of the sun. That is where I first learned about herons.

When a great blue heron family nested in a pine tree that overhangs her property in Orange County, California, Jean Macnab was pleased. When they returned with friends the following season, it was too much. She cleaned up several gallons of bird droppings from her patio each day, and then there were the half-eaten fish dropped from the treetops. Still, it was only when development threatened the rookery tree that someone did something about the problem. After the season's families left, a research biologist moved the empty nests to an established rookery to entice new residents to move in.

Herons are wading birds. Generally, they have a long neck and legs and a sharp beak for fishing and digging in the mud of shallow waters. Herons are carnivorous and may regard your fish pond as their own. Centuries ago, the Japanese used a *sozu kakehi* to frighten away wild boars and other animals from gardens. A *sozu kakehi* is a bamboo-pipe off-center see-saw. A gentle stream of water flows into the shorter upper end of the pipe and fills it. The now-heavy upper end swings down, empties its water, and abruptly pivots upward, banging the lower end of pipe on a strategically placed stone. The diaphragms between bamboo sections prevents the water from leaking from the upper to lower cavities. The slower the upper part fills, the better.

To construct a *sozu kakehi* you need a bamboo pipe about forty-two inches long. Glue the pipe to its dowel axle, which will turn in the holes of the two supports. To make sure you get enough weight to tip the empty pipe downward again, you must position the axle about two inches from the middle toward the upper end of the pipe. Make sure the water supply doesn't touch the open pipe. Larry Manger, wildlife biologist with the U.S. Department of Agriculture Animal Damage Control Department recommends the bamboo contraption as being "really good for keeping herons out."

"This one gal in particular who has one of the most beautiful koi ponds I've ever seen installed one. She has not had a heron problem since. Herons can be a real problem. That's something the terra-cotta pipe will give you a lot of protection on, because sometimes they'll sit there for hours, just waiting, and waiting and waiting, and when the fish does finally get into range, they're lightning fast and the fish is gone."

Herons are considered migratory birds, and it's illegal to hunt, capture, kill, or possess them.


## CROWS

The common crow is magnificent. It's a large, blue-black bird, reaching lengths of almost two feet. Their caw-caw cries are known to both city and country mice throughout the country.

Crows are known for their intelligence. I actually saw one perched on a newspaper vending machine on upper Connecticut Avenue in Washington, D.C.—and he seemed to be reading the headlines of the papers. Often people keep crows as pets, enjoying their antics. Behaviorists have trained crows to solve simple puzzles; they can mimic human speech like parrots.

Crows are omnivorous, eating just about anything as shown by studies that have counted over six hundred foods. Crows are indisputably more numerous than they were before the arrival of white settlers on these shores, and in some places today they gather in roosts of more than a half-million birds. In addition to a love of grains, they eat a number of harmful insects like grasshoppers and cutworms. Insects, carrion, eggs, reptiles, fish, and young birds make up about a third of the crow's diet. The rest comes from vegetable matter, crops and wild, and other miscellaneous items.

Crows breed in the early spring, forming pairs to build a coarse twig and stick nest, which is lined with softer materials like feathers, grass, and cloth. Nests are usually fifteen or twenty feet high in trees, or maybe on a telephone pole. If trees or poles aren't available, crows may nest on the ground. Males and females share incubation of eggs and care of the young. The female lays four to six eggs that hatch after eighteen days. The young crows leave the nest after five weeks to feed with the parents, and they stay together as a family unit for the summer. Often a family member stays high in the trees to act as sentinel while the others feed.

As fall looms, the families join other families to form larger groups, and larger groups gather to form immense congregations, so that by late winter, in some areas, the crows number in the millions. They disperse during the day to feed, but return at night to be with their kindred. Crows live about five or six years in the wild, though some may live as long as fourteen years.

Use noisemakers and visual frightening devices to discourage crows from holding their winter family reunions too close to your house. Gas exploders, recordings of distress calls, mylar ribbon or balloons, alarms or bright lights may scare the birds away. Vary the volume and location of the surprises so the birds never know quite what to expect.

Most pest control experts do not think of crows as a problem for homeowners. Becca Schad says "I've had calls from people complaining about crows eating from their birdfeeders, but it ended up that the people were putting out corn [which crows love]. And I thought, Well, you're feeding the birds . . . and you're getting results."

If you are bothered by crows at your feeders, try setting up a separate feeding area for them and spread cracked corn for them to eat. If they bother your garden, scattering corn in a selected area may also lure them away from your seedlings.

# WOODPECKERS

Woodpeckers are well adapted to their roles as the jackhammer of the animal world. Their hard, sharp beaks break holes in the wood so their long tongues can grab insects inside. Woodpeckers can balance easily on the side of a tree because they have two toes facing backwards and a stiff tail for support.

Twenty-two species of woodpecker grace the United States and are most likely found in wooded areas, where they find food and shelter. Woodpeckers have adapted somewhat to development and find good eats and a place to live in telephone poles, fences, and other wooden structures. In addition to boring insects found in wood, some woodpeckers eat insects found on the ground, like ants, as well as berries, nuts, fruit, and some seeds. The sapsucker eats sap, as the name implies, in addition to insects.

Not all pecking results in a woodpecker gourmet treat. Some species poke holes in trees and poles just to store acorns in the little holes. Other species enlarge a hole for a suitable nest cavity, or create a cavity in a solid tree. Sometimes they're drumming to the beat of some internal music, the music of springtime, mates, and eggs.

Betsy Webb, curator of the urban wildlife of the Denver Museum of Natural History, has much experience with woodpeckers: "In the spring there is a particular problem with flickers, a type of woodpecker. They drum on the cedar siding." The male flicker in spring, during the breeding season, will drum on wooden siding, and it reverberates loudly both establishing a territory and advertising the male's availability. The louder they drum, and the more the wood reverberates, the better able they are to set up a territory and attract mates. Sometimes they'll actually excavate a nest hole in the side of a house, often in cedar since it's soft.

"In doing that, they sometimes cause damage to houses up in the thousands of dollars. They create a series of holes and the whole siding will have to be replaced, and that's very expensive," says Webb.

Try these series of solutions for dealing with the birds:

1. Get two one-square-foot pieces of wood matching the type of siding on the house. Nail them together at one corner so that they flap together. Then secure the back piece to the wall where the damage occurred. The flick flicker will focus his attention on the flapping piece of wood, which reverberates even louder. And you can set up a series of them to try to concentrate the flicker to do damage on

the wood squares. You just remove them after the breeding season.

What about the noise? Webb explains, "When you build a house in deer or flicker territory, you are going to be living with those species, and you can either have a cycle of negative problems with them year after year or you can put in place very creative solutions to live in harmony with them."

If woodpeckers are beating on a metal chimney, try wrapping it in burlap to muffle the sound. (Take down the burlap, though, before you start a fire!) They might get frustrated and find another boomer. Or try a lure—put up a metal tin some distance away from the house.

**2.** Hang high-quality images of birds of prey. Use flat silhouettes, like hawks, or the plastic owls. They need to be a thick reproduction of a substantial, threatening-looking predator. If it's placed before the breeding season starts, it is likely to warn off a bird. If you put it up after the season has already started, it is not as likely to change the behavior.

Birds aren't stupid, they're well adapted to their environment and are extremely opportunistic. Anything you can do to fool them must be done well. Move the silhouettes or predator reproductions occasionally so that the flicker doesn't get used to them. Try banging every morning with a pot and pan, but you'll have to do it every morning at five.

**3.** Attract them to trees in your yard by putting a flicker nesting box up in your yard. Your local Audubon chapter will have good suggestions for the specifications for flickers or for your local wood-pecker population. Birds have specific nesting requirements; they'll select a box that fits their specifications for entrance hole, depth, dimensions, and placement on the tree. These species are particular and just any bird-nesting box won't do.

In many parts of the country, it's a tight housing market for the birds. Mark Westall, a naturalist on Sanibel Island, Florida, says, "People build big wooden houses, clear away all the vegetation, take out all the palm trees, put in a sod lawn, then they can't understand why the woodpeckers are poking away on the side of their house.

"Lots of times a woodpecker will poke a big hole in the side of the house because he's looking for a nesting cavity, and what you can do is put a nest box right over the hole where he's working." If you give the woodpecker what it wants, you'll save your house. Nesting woodpeckers will usually chase away other woodpeckers, but you may have to put a nest box up on both sides of your house, because one pair may nest on one side and another on the other side. The wood siding survives and you get to enjoy watching the woodpeckers raise their young.

Don't be so quick to remove dead trees from your yard; they will give the woodpeckers a place to live and maybe feed. It works for Westall: "We don't have a lawn, we have pruning shears. We prune the jungle away from the side of the house and we can watch the pileated woodpeckers working on trees ten feet away from the side of the house. We don't have any problems with them working on the house because we still have plenty of habitat here for them." Friends of his even acquired a couple of dead cabbage palms to put on their property so they could attract woodpeckers.

But frustrated homeowners should beware before they do anything drastic: Woodpeckers are classified as migratory songbirds, and it is illegal to harm them. Also, if your house is wood, it's possible the woodpecker is doing you a favor—your home could have a termite or bug infestation you don't know about.

Keep in mind that often woodpeckers are interested in getting at the bugs in your house, not in making little holes in your house per se. Thus, getting rid of the insects that may have burrowed into the side of your house is one of the best steps you can take to deal with woodpeckers.

Westall says, "They make long strips up and down where they poke away the wood—usually on the corners of the house where the paneling comes together, and builders put trim over it to cover it up. Ants like to go up into your attic and follow that trail underneath that trim." The woodpeckers know that the food is there and they start cutting away at those strips. If you want to stop woodpecker damage, you have to stop the ants, and solutions range from boric acid in their nests to a professional exterminator to mothballs in the attic.

Westall's advice is applicable to dealing with all animals, including woodpeckers: "The main thing is to think like wildlife. Say to yourself, 'That woodpecker isn't damaging my house because he doesn't like me; he's doing it because that house gives him something he's looking for. Now am I smart enough to figure out how to give that to him without causing damage to my house?' "

# GENERAL OBSERVATIONS

If you have a garden you will have noticed that you have created one very large bird feeder. That's okay, if that was your intention, but most people don't envision this as the main purpose behind their garden. Birds have an advantage over all other animals: They can fly. That means fences are extraordinarily useless.

But bird netting is not. A large garden may make netting difficult, but a smallish garden can support it very well. Netting will reduce bird infestation nearly perfectly.

Mark Fenton, president of Peaceful Gardens in California recommends, "Row covers are good bird and insect protection. The barrier method works. There are some bird repellents, the Savanna bird repellent, for example, or you can make one at home with Vaseline and cayenne pepper. I think the scare-eye balloons are one of the more effective in bird control. [Scare eyes are scary faces.] I've used them personally. The birds get pretty skitterish; it won't stop them completely, but they fly in and peck something, see that thing and fly out again. But they won't stay. The main thing if you have a bird problem . . . is to use various different bird things all the time. In my own orchard, I put up some scare eyes and mylar tape this year and have way, way better control than I had last year with nothing there. Last year they got all my apples before they ripened."

While large nets are effective in keeping birds away from the fruit, they are sometimes impractical. Al Geis, research director of the Wild Bird Centers of America, suggests that often an alternative food source will distract birds away from your crops. "A food they like just as well that fruits about the same time is tartarian honeysuckle, an exotic plant. Most of our native and even our ornamental fruit-bearing plants fruit too late. The fruit doesn't become available when strawberries or blueberries are available." While the tartarian honeysuckle does not spread aggressively like other exotics (the notorious multiflora rose, kudzu, and Japanese honeysuckle), you may want to check with your local USDA extension office before you plant it.

An exploder could be equally effective against small intruders. Fenton explains that the birds won't grow accustomed to the sound "because they never really stay around. The birds just don't live right next to it. They'll fly in and get scared away." The noisemakers should be placed in the middle of fields or orchards, far away from civilization, and allowed to boom all day long. The device doesn't shoot out flames, so it can be left unattended.

Of course bird problems don't stop at the bird-feeder or garden. They can come right into your living room, so don't underestimate the importance of excluding birds from your home using chimney cages and screening over vents. A Los Angeles family was stunned when hundreds of migrating Vaux's swifts swooped down their chimney and into the house. The swifts, on migration from Central America to the Pacific Northwest became confused when they mistook the chimney for a sheltered resting place and then ended up in a strange living room. They spread soot and droppings throughout the house before exiting through open windows and doors. Dozens died in the attempt.

Check the resources section for all sorts of implements for frightening and excluding birds.

## ❖ BAFFLING BIRDS

1. Exclude them from flat roosting surfaces with porcupine wire, netting, a sloped ledge, or a tacky, irritating substance.

2. Cap chimneys and vents so they can't enter the house; make sure the attic is secure so they can't get in there to nest either.

3. Buy special bird-feeders or birdhouses to attract only one type of bird.

4. Offer alternative foods to your garden.

# RACCOONS

Practically anyone who's gone camping has come back with an "Aren't They Clever !@#$!" story about raccoons in the wild. For example, during a camping trip in West Virginia, a friend of mine sealed up her perishables in a plastic bag and submerged the bag in the river overnight to keep cool. In the morning, she found that raccoons had lifted the bag out of the water, untied the knot, removed and opened the carton of eggs, neatly broken each egg and sucked out the contents. Then they'd carefully placed the eggshells back into their individual cups and closed the carton.

That's almost enough to make you forgive them. But raccoon antics in suburbs and cities—where they're increasingly making their homes—aren't so endearing.

Larry Manger, wildlife biologist with the U.S. Department of Agriculture's Animal Damage Control Department in California, tells the story of a California man who owned a beautiful Porsche. Raccoons got into his garage one night, and in the course of some mad rummaging, managed to knock several heavy filing cabinets over onto the Porsche. Manger says they did ten thousand dollars' worth of damage to the car.

Then there was another California fellow, who lived in a large, beautiful home, four or five thousand square feet in size. Apparently he spent most of his time at one end of the house, so he didn't realize that raccoons had torn off part of the roof on the other side of the house. Raccoons sometimes do this sort of thing when they have young, Manger says. "If they can't get in through a screen or something, sometimes they'll just get an inkling they want in, and they'll just rip a hole in the roof and go in." Well, one night the man was watching TV in the raccoon's side of the house. It rained rather heavily, the plaster got soaked, and the whole ceiling came down. "He ended up with close to twenty-five thousand dollars' worth of damage," says Manger.

It's sometimes hard to believe an animal of this size can cause such havoc. The raccoon is a small mammal, typically twenty to thirty inches long, weighing fifteen to thirty-five pounds; though in urban areas—where coons thrive on our refuse—they can weigh up to sixty pounds. Their fur is a grayish brown, but what every school child remembers about a raccoon is its bushy, banded tail and black mask.

There is an appropriateness to the raccoon's banditlike appearance. They're smart, curious, bold, fast, and sneaky; and they will snoop into anything that smells like food—no matter how well secured. And they've got the kind of clever fingers that some safecrackers would kill for.

Their front paws are actually shaped like little hands. (In fact, you can tell a raccoon's been around, because the prints of their forepaws actually look just like tiny baby hands.) They have five dexterous "fingers" that they use to pull, pry, turn, twist, grab, and rip their way into all sorts of trouble. And they have the persistence and—despite their size—the strength, to go after whatever they put their devious minds to.

While many animals suffer when humans develop the land, raccoons seem to thrive. They live well in woods, suburbs, and cities. Raccoons range from southern Canada to South America. In the United States, raccoon populations are estimated to be at about the same levels today that they were in the mid-1800s.

The reason for this success is that coons are so adaptable. Take their diet, for instance: They're omnivores. In laymen's terms, that means they eat almost anything. The raccoon eats fish, turtles and their eggs, shellfish, salamanders, and insects, bird eggs and birdlings (they can decimate a waterfowl nesting site), young mammals, fruit, nuts (especially acorns), grains, almost anything growing in your garden (especially sweet corn), cultivated crops, garbage, and other food associated with humans—pet food, scraps around the barbecue, and bird seed. They'll even brave wasps' and hornets' nests to get to the goodies inside. They seek a variety of foods, and they savor crayfish and frogs. As suitable habitat disappears, they seek out our gardens, yards, and garbage for food sources, but even then they vary their food sources.

They also eat pizza, which may be the real reason they are thriving in suburbia. Witness what happened to one homeowner one evening when he was having dinner outdoors on his patio. The man stepped inside for a couple of minutes to answer the phone, and when he returned to his meal, "the unwatched pizza was in the middle of the backyard," with a raccoon munching on the mushrooms, pepperoni, and cheese. He even liked the anchovies.

Raccoons are known to dunk their food in water before consuming it, a habit that earned them the name *Procyon lotor* from the early biologists. *Lotor* means "one who washes." The genus name was made up, because no one could figure out what family the raccoon belonged to—at first Linnaeus thought it was a bear. Anyway, no one knows why they wash the food. Some speculate it has to do with making sure live food is really dead by drowning it if it's not. Maybe it makes the food easier to eat—kind of like dunking a stale donut—except it's turned into a kind of nervous habit. It might just taste good to them.

In suburban settings, raccoons have sometimes had to give up this fastidiousness for lack of a ready water source (although when eating pizza, perhaps they dunk it in Pepsi). No one really knows why they dunk their food; it's certainly not to clean it, because the food is usually consumed regardless of its condition. They have definite tastes. Author and naturalist Virginia Holmgren has watched and fed raccoons in her backyard for over thirty years and found that the animals would filter through bakery scraps to first eat the sweet breads and eat rye and pumpernickel last.

Coons are also adaptable when it comes to their accommodations. Good climbers, they often prefer to den in a hollow tree. But, they'll also intrude on other animals, using burrows and dens others have created. In other cases, they may simply use a secure, sheltered rock crevice, tree branch, or squirrel's nest. They also like our buildings, sheds, and sewers as denning sites.

Because the raccoon is nocturnal, you rarely know it's out there unless it does some damage to your property or unless you happen to spy one during the daytime as it naps and suns in a treetop or rock crevice. (You also might overhear them chatting as they munch on your garbage: They vocalize with a soft churring sound when they feed.)

Because they rarely stay successive days in the same den, raccoons keep several sleeping dens in their home range, which can be as small as ten acres and as large as several square miles. The size depends on how much food and how many animals there are in any given area.

Home ranges can overlap, and the raccoons develop a social hierarchy to allow their coexistence. A male controls a larger area than a female, and several females may live in his territory. The first meeting between strangers probably involves a fierce display and posturing, and even fighting. This establishes who's boss, and at subsequent meetings, the loser will give way. They may scent-mark their territory by leaving scat as a marker or by rubbing their anal scent glands on objects.

In the cold North, raccoons den up for the winter, but in the South they remain active. Northern raccoons spend the summer eating and putting on weight for winter, when food is scarce. They may double their weight, not too difficult a task considering their catholic tastes. Still, they don't hibernate, and have to remain somewhat active over the winter. Early in the century, Russia imported raccoons in an attempt to raise them for fur. Unfortunately, they didn't do so well over the cold winters and many died. Survivors made their way to Germany and survive there still.

Raccoon fur has always been popular with raccoons, but people have coveted it too. By the 1920s, raccoons were in danger of extinction because of a faddish demand for raccoon coats; it took fifteen skins for each coat. Raccoons have bounced back since then, with only a little help from us.

Raccoons are mostly solitary except in mating season and when they have young (most of the year). A breeding-age female is rarely alone, perhaps from the time the last kit of one litter leaves until a male arrives to mate to produce the next litter. Adult females may occasionally hunt together. Mating season begins in December in the South and in January or February in the North.

Females mate when they are a year old; males wait until they are two, living for a few years alone or in bachelor groups. The male is polygamous and will mate with and live with a female for several days or weeks before going off in search of a new mate. If the female does not become pregnant after mating, she will become fertile and mate again. A den for the family has to be really secure, so the female will seek a suitable site, one with a protected opening, preferably high in a tree.

After breeding, the female stays alone. About sixty-three days after conception, she gives birth to three or four kits, helpless, four-inch-long, two- or three-ounce blind animals. They have a little bit of fur, but by the time they are ten days old, they sport the characteristic facial and tail markings. The mother stays with them for a few days more, but then resumes her nightly foraging for short periods, staying close to the den.

After about three weeks, the kits' eyes are open, but they sleep most of the time. By the age of one month, they weigh about two pounds; the mother now leaves them alone all night while she forages.

If the den is high up, at this stage, the female often moves the offspring to a different den, one that's lower to the ground, because as they begin to explore the kits will sometimes fall out of the nest. While the coons can survive drops of up to twenty feet, mother raccoons seem to prefer not to put this to the test so early in life.

The little coons can walk, run a little, and climb when they are two months old. When they are about ten weeks old, they begin to get weaned and start eating solid food. Pretty soon they roam far and wide with their mother and begin to sleep in different dens in the home range. They are completely weaned when they are twelve to sixteen weeks old, but they still feed in the mother's home range, even if they don't come back to her den every night. In the North, offspring may spend the winter in the same den with the mother until she needs the space for a new family and drives them out. Or, they may just den nearby. In the South, the young seek their own way in the fall, with females generally staying nearby and males seeking new territories.

As long as the temperature stays above twenty-eight degrees Fahrenheit raccoons forage, but they sleep the night away if it is colder than that. As the winter passes, the animals may grow acclimated to the cold and will be out when it is as cold as zero degrees. At any rate, by the springtime, they usually weigh about half what they did going into the winter, and late winter and early spring are especially hard on the animals.

Raccoons born late in the season may not have had time to build up fat stores to survive a winter; they may starve or become prey to another animal. Adults, however, are fierce opponents and are rarely prey, except to human hunters and cars. All in all, they probably live around six years in the wild.

## Raccoon Encounters

In suburban areas, towns and cities, raccoons may live in parks, in a patch of woods, near streams, or even in your backyard if it provides suitable habitat. Suitable habitat means access to food and water and a place to live—a denning tree, a sewer or hole underneath your house or other buildings, an uncapped chimney, or a cozy attic. Unsuitable habitat, to a raccoon, is a yard containing a dog.

Although it may be exciting, initially, to discover that you have raccoons as neighbors, their penchant for overturning your garbage can, tearing holes in your roof, and terrorizing your birds can quickly temper your enthusiasm.

Rabies is also an issue of growing concern. The eastern mid-Atlantic states have seen a rise in the number of raccoons carrying rabies. The virus moves through the country like a wave, peaking and receding. For example, in 1981 in Maryland there were only seven reported cases of rabid raccoons; in 1984 the number had risen to 964. It's even in Westchester County, near New York City, with seventeen reported cases in less than a year. To test an animal for rabies, it first must be killed, and anyone bitten by an animal who can't be tested is treated for rabies with a series of five shots.

For this reason, raccoons and other wild animals are dangerous.

(Skunks, foxes, groundhogs, and squirrels may carry the virus as well.) Raccoons generally don't attack people. Rabid raccoons, however, are an exception. A rabid animal may be especially friendly, lethargic, or disoriented, but lethargy is also a sign of distemper, a disease not contagious to humans. Report any oddly behaving animals to wildlife authorities.

Raccoons also carry an organism called raccoon roundworm, which can infect people who come into contact with raccoon feces—while walking in the yard barefoot, for example—or who inhale the roundworm eggs, which, however, is unlikely since raccoons, like cats, are private about their toilet habits.

Another caution about raccoons: Female raccoons that are not rabid may attack people if they perceive a threat to their babies. So in the summertime, be particularly careful around raccoons.

## In Your Yard

### In the Garbage

If raccoons live in your vicinity, chances are you'll know it because they will have made a raid on your garbage cans. Raccoons love trash and are regular Houdinis at breaking through whatever systems you devise to keep your cans closed. Raccoons usually knock over a garbage can to get at your tasty leftovers, so if you can keep your cans upright, you may begin to thwart the masked marauders. To keep them upright, place your cans in a rack or tie them to a support, suggests the USDA extension service. But it can take some doing to devise a lid they can't conquer.

I used to store birdseed in a metal garbage can in the backyard. Although raccoons don't really care much for birdseed, and aren't attracted by its smell, periodically they would take the can's top off and rummage through the seed bags, just to see what was inside. Before I became more sophisticated about outwitting critters, I closed the can with bungy cords. Of course, they ate right through the cords. From then on I kept the seed indoors.

Another way to tackle this problem is to make your garbage less appetizing. You can buy repellents like Ropel, a sticky, foul-tasting concoction that you spray onto the outside surface of your plastic garbage bags. One taste and the raccoons will seek dinner elsewhere.

Cal Saulnier, with Plow and Hearth Catalog, can personally vouch for how foul this stuff is. "I've tasted it, and it's disgusting," he says. Other pluses: It's relatively inexpensive and harmless to animals and the environment.

Still another approach involves making the cans less inviting. Several catalogs sell a small mat that delivers a mild electric shock to animals that step on the surface. These mats, which go by names like the Scat Mat or Invisible Gate, are supposed to be placed under the cans. The theory is that the negative stimulus will train an animal to avoid these areas. Of course, you may have trouble getting anyone to take out the garbage. The device usually sells for around sixty dollars, and if it fails outside, you can use it to train your indoor pets to stay off the couch.

You can also try to make your garbage cans less accessible. Try putting the cans in a place where the coons can't get at them, like in a sealed garage. A related strategy I've seen recommended is to put your can inside another, larger can. I don't know if this has actually ever worked, though. It strikes me that raccoons might view this simply as added challenge and entertainment—a raccoon version of those hollow Russian dolls that each have a progressively smaller identical doll inside.

Plain old bribery may also work. Some California residents found that by offering pet food, they were able to persuade the raccoons to leave the garden and lawn alone. Raccoons in the yard take some getting used to; they still may attack the lawn, garbage, and garden—but not with the hunger of before—plus they'll leave footprints and evidence of play all over the yard.

### On Your Lawn

Raccoons like the grub worms that infest thick lawns. In a well-established lawn, they'll dig for the creatures. And, if you've just laid down new sod, watch out. Raccoons may roll back your new sod searching for tasty grubs.

To keep the animals off your lawn, there's an old remedy that calls for spraying the lawn with a mixture of one cup each of children's shampoo and ammonia. It should keep the raccoons away for a season. That formula has been updated by substituting one cup of Hinder animal repellent for the ammonia. You can obtain Hinder through a farm supply store or catalog.

You could also apply milky spore to the lawn. It will parasitize and kill the grubs, so the raccoons won't have any reason to dig up your lawn. However, milky spore takes years to get going, so you'll need a short-term solution. Try altering the watering of the lawn. The grubs can't thrive in dry soil.

If you're trying to save your sod, one solution—assuming you're not dealing with too large an area—is to pin the sod down with stakes or wire pins.

### In Your Fruit Trees

Fruit trees are another raccoon temptation, and excluding them from these forbidden fruits can be a problem. But *if* the tree is isolated enough so that the animal can't jump into its fruit-laden branches from a nearby tree or house, there is hope.

One way to tackle the problem is to "flash" the tree trunk: put a collar of slippery metal around the trunk, which prevents animals (squirrels, as well as raccoons) from climbing up the tree. Metal flashing is thin aluminum, which comes rolled on a spool. Look for flashing that is at least three feet wide; that way, it'll stick out far enough from the tree that the animal can't reach out to the edge and pull itself over it. Tack the metal around the tree, being careful not to make it so tight that you will girdle the tree.

### In Your Pond

In the northern part of Sacramento County where Larry Manger works, lots of people have installed small ponds for koi, expensive oriental carp. Naturally, the wildlife enjoy dining at these pricey piscaries. Says Manger, "I've had people with koi worth thousands of dollars, ones they've imported from overseas; they come out in the morning and find the raccoons have eaten them." Fishing birds like herons and kingfishers also eat fish from backyard ponds.

One way to protect fish ponds is to make them more than two

and a half feet deep, Manger says. The way the raccoons catch back-yard fish is to wade into the pond, corral the fish into a corner, and grab them. But in deeper ponds, where raccoons have to swim, they can't do that and the fish can escape by swimming underneath the coons.

If you don't want to deepen your pond, one inexpensive and effective strategy is to place some pieces of terra-cotta pipe, about sixteen or eighteen inches long and four inches in diameter, on the bottom of the pond. This gives the fish someplace to go when the raccoon is in there swishing around. The fish can go in the pipe, and when the raccoon sticks his hands in one side, the fish can swim down to the other end. This gives the fish enough cover that they can maneuver and have a better chance of getting away, he says. Also, the pipes won't detract from the aesthetics of your pond. "After they're in the pond for a while, the pipes moss over and you can't really see them," he says.

According to Manger, the most effective approach is to install a small electric fence around the pond. Manger thinks these protect small ponds from raccoons, but he qualifies his approval by pointing out that "an electric fence kind of detracts from the pond."

### In Your Birdhouse

Another target for raccoons is your birdhouse. Raccoons find bird eggs a delicacy. Here's where technology helps. The Bird Guardian is an effective ploy for outsmarting egg- and birdling-eating critters. The simple mechanical device adds a three-inch-long tunnel to the open-ing of a birdhouse. A hungry raccoon or cat won't be able to simply reach its paw into the house opening to snatch up helpless nestlings.

However, some birds, most notably some bluebirds, are frightened by the device and won't enter the birdhouse through it. To address that, Bird Guardian has recently added a short ladder outside the tunnel, which the bluebirds can use for balance; perhaps the new model will be more attractive to them. (While the Guardian will stop a short-armed mammal, it won't halt a snake, so you'll need other measures for snake problems. The Eastern black snake, for example, is an adept climber.) While the Bird Guardian is an effective tool, a determined raccoon (is there any other kind?) may be able to pry it off.

## Denning in Your Trees

One reason you may be getting so many unwelcome visitors to your garbage, lawn, or birdhouse is that the raccoons may actually be denning in your yard.

If that's the case, one way to encourage them to move is to make your yard an uncomfortable place for them to live.

As I mentioned earlier, getting a dog is one approach.

Another method is rock and roll. Raccoons, like many small mammals, seem to have an aversion to loud rock and roll music—heavy metal in particular. An outdoor loud speaker aimed at the raccoons' nesting place should get rid of the raccoons that once considered your place a peaceful haven. The music might attract neighborhood teenagers, however; but that's a tradeoff you'll have to make.

Though expressed in a tongue-and-cheek way, this is a principle you can keep in mind for most critters: Unnatural noises scare animals. The louder the better. The stranger the better. If raccoons are a persistent problem, you might consider the following apparatus: An outdoor speaker connected to an infrared motion detector. When anything larger than a house mouse approaches, a loud noise goes off. Works wonderfully.

## On Your Deck

My own raccoon problem involved a group of raccoons that was using an unfinished deck off the third floor of our house as a rest and play area. This deck was surrounded by several trees, so the raccoons could easily visit it.

At first I thought about feeding the raccoons—after all, they are awfully cute and they were *outside*. But my wife, Peggy, advised that once I gave them some handouts there would be nothing I could ever do to get rid of them. In addition, there was the possibility of rabies among the raccoons; a growing problem in the mid-Atlantic states.

So, I developed a simple, ultimately nontoxic solution: I spread several dozen mothballs on the deck. I would never use mothballs inside, where harmful levels of naphthalene can accumulate, but outside, the fumes would waft harmlessly away. Not only do moths have an aversion to mothballs, but raccoons aren't crazy about their smell either.

In the month it took for the balls to dissolve into the air, the raccoons discovered some other place to go. Raccoons don't like most strong, sharp smells; dousing the deck with ammonia might have been just as effective.

## In Your House

There are two reasons raccoons come into your house. Some coons come in just for temporary visits; others are trying to set up permanent residence. There are different ways to deal with each kind of intrusion.

### In Your Living Areas

Some more brazen raccoons just march right in through pet doors, usually because they smell pet food inside and want a snack.

Betsy Webb, curator of the urban wildlife exhibit at the Denver Museum of Natural History, had a raccoon that would enter her home nightly through a dog door. While she maintains "My mission in life is not to keep animals out of my house," this particular mother raccoon repeatedly helped herself to Webb's food.

At first, Webb used the dog door's wooden slats to secure the opening at night, but the raccoon punched the door open. Then, Webb tried propping a chair against the small door, but the raccoon pushed into the home for its midnight kitchen raid. Webb finally had to close off the door entirely until the raccoon got unhabituated to entering the house and eating.

Webb removed the barrier after a few months and had no more problems. She thinks the female raccoon was entering the house during times when she had young and needed the extra food.

If you come across a raccoon in your kitchen, the best tactic is to open your doors and windows, back away, and let the raccoon leave of its own accord. If you try this technique at night, turn off the lights in your house so that you don't invite every moth in the neighborhood inside.

Once the coon leaves, secure the pet door.

People also occasionally encounter raccoons in other rooms. Marta Vogel, of Takoma Park, Maryland, had raccoons that apparently came in through a heating vent, made their way down to the furnace, then "by squishing their bodies," actually came out through the furnace and into the basement.

Brazen creatures, they didn't stop there. Sometimes they even wandered upstairs. Marta's roommate Valerie occasionally mistook the coons for one of Marta's cats "because our cats have stripped tails." What did Marta do about the problem? She moved. "I left them for the next tenant."

If you come across a coon inside your house—and don't want to take Vogel's laissez-faire approach—"Remember first that the trapped raccoon is as frightened as you are," says Laurie Bingaman, with the National Zoo. Then act, moving slowly and gracefully. Isolate the intruder in a room that has access to the outside, then open windows and doors so the animal can escape. Becca Schad, owner of Wildlife Matters, an integrated pest management firm in Virginia adds, "Pretty much, when left to their own devices, they'll get themselves out. They realize it's not an appropriate spot" and will try to leave.

But if the animal seems too confused or afraid to act in its own interest, you may have to motivate it. Walk slowly toward the critter to herd it outside, but don't wave your arms or yell, which may confuse it. Before you begin the procedure, you may want to slowly and gently remove Grandma's teacups and saucers—or any other valuable breakables—from the room, since things may get exciting. It's also helpful to have a broom, not to flail around, but to use as an arm extension to block a movement in the wrong direction. As the animal exits, in the words of Bingaman, "lunge towards the open door or window and close it."

### In Your Chimney

Chimneys are warm, there's always heat escaping from them, and raccoons (not to mention other animals) are attracted to them (as long as there's no fire) for winter dens and great spots to raise a family. The chimney is a case where an ounce of prevention is worth a pound of cure. It's easy enough to raccoon-proof your chimney, and it's wise to do it if raccoons are ever seen anywhere in your vicinity. If you don't, you may well have problems.

Most hardware stores carry a cap that effectively blocks entry to your chimney. It is made out of wire, and once it is attached to the chimney top, it allows smoke to exit but won't allow animals to enter. Chimneys are different sizes, so before going to the store, you'll need to measure your chimney in order to get the right size. Check with your hardware store to see what's available.

Strong hardware cloth should work as well; just attach it securely to the chimney with masonry screws.

If the animal (or animals—a mother may give birth and nurse her young in a chimney) is already living in your chimney, the problem is a little more complicated. One way to tell if you have chimney residents is if you hear mice-like squeals coming from the area above your damper.

If you suspect you have residents in the chimney, start by blocking off their entrance to your house. If, heaven forbid, the flu is open, block the opening into the room. Otherwise, block the fireplace opening securely, keeping in mind that a mature raccoon may weigh thirty-five pounds or more and can move rather heavy objects.

If your resident is a solo adult, one option is to wait for the animal to leave on its nightly rounds and then scramble up on your roof to install your animal-proof chimney cap.

If your unwanted tenants include immature offspring, you can wait for them to grow old enough to leave on nightly rounds with their mother and then cap the chimney. The young nurse for two months, so if you take the wait-and-see strategy, you won't be waving goodbye to the family until the young are weaned.

However, if you can hear the babies, they're probably already at least six or eight weeks old, points out Becca Schad. Furthermore, that means they've already "been there for several weeks, so what's another couple of weeks?" she points out.

She recalls one client, a woman over eighty, who complained about noises in her chimney. "When I told her she had a raccoon family in her chimney, she said, 'Well, I guess I'll just have to wait.' I just wanted to hug her!"

If you don't have that kind of patience, you can try to encourage the animals to leave on their own. To do that, use "some mild form of harassment," to disturb the mother and make her aware that you're there, points out John Hadidian of the Center for Urban Wildlife of the National Park Service.

You can shine a mechanic's trouble light down the chimney and put a radio on in the fireplace; raccoons don't like bright light and loud noise and will generally move. Since they also don't like certain strong smells, you can try placing a cup of ammonia in the fireplace, or dropping some moth balls down the top onto the flue.

You can also call in a professional to remove them. But bear in mind that professionals will usually kill the animals.

### Under Your Eaves or Under Your Porch

Raccoons living in your attic (or under your porch) are best dealt with by leaving the lights on and keeping a radio going nearby. This will persuade the raccoons to leave.

But then your task, of course, is to find their point of entry. "There are always structural problems when an animal is in a house: a hole in a board that leads to the attic, a hole in the roof, an unscreened opening underneath a crawl space," says Hadidian.

Even if the animal seems to have left for good, it's important to go ahead with the repairs to prevent the same animal, or other ones (or weather) from getting in. "In ninety percent of the cases where people haven't made the repairs, I get a call the next year because the raccoons are back," Hadidian says.

And what if you can't find that point of entry? Now you get to play detective. Look around your house carefully to find the possible access points; watch them at dusk to see if or where an animal emerges for its nightly rounds. And don't ignore even very small holes. "If an animal can wiggle its head through a hole, it can usually get the rest of its body through," says the National Zoo's Laurie Bingaman.

Once you've zeroed in on the hole, wait until the coons have left on their rounds, and then plug it up. However, if you don't want to wait until after sundown to do your carpentry work, you can try attaching a one-way door at their point of entry. Hinge the door so that it swings out, but not in. Then, once the animal has left for good, you can close up the entry point permanently.

If you're about to block up a hole, take along a bright flashlight as a precaution. Not only will it help you as you work, but in the event that you find a raccoon inside the attic or under the porch, the brilliant light will temporarily blind it. That gives you that second or two you need to get the hell out of there. (Raccoons are not likely to charge, but bumping into one in a small, dark place isn't advised.)

It's important to keep in mind that blocking a raccoon's entrance into your home—or to its den in your yard—does not always have magical effects. A raccoon establishes several dens in its home range. Though you may have kicked him out of your house, he will likely continue to live in the area. And he may well show up again in your yard. Sometimes, "it seems like once they've decided they want to make your place home, not a lot works, except maybe a big dog," says Larry Manger.

If you have an animal that stubbornly refuses to leave, or that is causing particular damage, you may want to consider trapping. "Trapping is a last-ditch effort," Manger says. However, it can solve the problem. "Often you have a particular animal with a particular habit that's causing the damage. Lots of times you only have to reroute one animal." For example, he says he has been called in on several cases where raccoons were killing chickens. "I'll take one raccoon out, and that will be the end of it, even though there are still other ones in the area. The particular one that had the habit is gone."

Wildlife experts recommend not trapping the animal yourself unless you know what you're doing. In some states trapping may be illegal, and in all states it is dangerous because raccoons can give a nasty bite and, as mentioned before, often carry rabies.

If you have an animal in your house or yard that you suspect is rabid, call animal control experts immediately. They will trap and kill the animal.

If a problem raccoon does not appear to have rabies, the wildlife professional may decide to release it after it's been trapped. Be sure that the animal will be transported a good distance away, about twenty miles. Otherwise, these nomadic creatures may make their way back to your area.

One final consideration: The mortality rate for raccoons that have been moved is very high. It is estimated that nearly fifty percent die in the first three months after the move. With that statistic in mind, you might want to be sure that you've first tried to get the animals to move away on their own, and consider calling in someone to trap the raccoon only after other methods have failed.

## ❖ RULING OUT RACCOONS

1. Keep the garbage stored in an outbuilding or make sure raccoons won't be able to get the lids off.

2. Treat garbage bags with a foul-tasting repellent.

3. Make your home secure against nesting and denning raccoons.

4. Cap your chimneys.

5. Don't handle any animals in your house; let them find the door themselves.

# SKUNKS

There is almost nothing as pungent as a skunk. I don't care how long your college roommate went without a shower. Ask the residents of Gilroy, California, which boasts that it is the "Garlic Capital of the World."

With its annual garlic festival, Gilroy thought it had devised the smelliest event imaginable. But skunks managed to upstage those puny bulbs. While feasting on some garbage cans at the festival grounds during the 1990 festival, it seems that some skunks got riled and decided to compete for the pungency title. When they added their own perfume to the air, even seasoned garlic sniffers were forced to scatter, and the event was nearly ruined.

Incidents such as these have given the skunk a public relations problem. Year after year, these critters show up on America's least-favorite-animal list. This causes skunks some consternation, since they are basically timid animals who do not go looking for trouble. And, unlike some other mammals, they don't run around displaying large teeth, sharp claws, or unusual strength. In fact, ranging from six to fourteen pounds and measuring twenty-four to thirty inches (one-third of which is tail), you might even say that skunks are rather puny.

No, their public relations problem basically comes down to one of odor: A decidedly unsociable odor. And, like so many difficulties that arise in both the human and animal world, these troubles can basically be traced back to a pair of glands.

In the case of skunks, these are scent glands, located on either side of the anus. These glands secrete a wickedly disgusting liquid, a sulfur-alcohol compound, which the skunk sprays on perceived attackers.

Interestingly, skunks apparently don't like the smell of this either. They prefer to flee and save the chemical onslaught only for threats they can't elude. A skunk may warn the intruder by stamping its front legs and shuffling backwards, and it may also hiss and growl, before resorting to spraying.

But it's probably too late when the skunk forms its body into a U-shape, keeping all four feet on the ground, with its face and up-raised tail facing the target. The oily liquid shoots fifteen feet from one or both of the scent glands. One variety of skunk, the spotted skunk, sprays over its head while balanced on its front legs.

This proves to be quite effective, because the skunk is rarely any-one's dinner. And, once other animals tangle with a skunk, they generally don't forget the lesson. (The exception, naturally, is your dog. But more on that later.)

The skunk ranges throughout the United States (except in deserts) and in southern Canada. The skunk lives in many different environments from prairies and woods, to farms, suburbs, and even cities. The more familiar and more common striped skunk is found nationwide except in the range of the spotted skunk, found in the Delaware-Maryland-Virginia tristate area and from New England west through New York and Wisconsin and south to Illinois. The skunk family also includes two other members: the hooded skunk and the hog-nosed skunk, found in the extreme southwestern parts of the Great Plains. However, many of these skunks' behaviors are similar.

Skunks are solitary creatures; and other than a mother and her offspring, you rarely see skunks together. They are nocturnal and search for food at night, traveling a mile or two, though they never stray far from water. They are opportunistic feeders and eat a variety of plants, animals, and insects that they encounter. Of course, during the summer that means they eat a lot of insects—grasshoppers, beetles, crickets, and grubs. They'll also eat mice and even larger mammals like rats and rabbits if they can get them.

On the ecology scorecard, skunks are clearly in the plus column. This is because they have a strong appetite for insects as well as mice. Their diet also includes shrews, rabbits, bird and turtle eggs, berries, apples, grapes, and, of course, garbage. Sometimes they raid chicken houses and feast on eggs.

In warm climates or seasons, skunks spend their days resting in grassy fields, brush, or other places that afford good cover. They are most active at night and may travel a mile or two searching for food. In colder weather, they seek out a good spot to keep warm—often a burrow or hollow log.

Skunks like to do as little work as possible on their winter dens. Although they may dig their own holes, usually they'll use an existing woodchuck hole or perhaps a crawlspace under a house. It's not unusual for the solitary skunk to gather with other skunks in a winter den; usually a male will share a den with one or more females. Once in a while, they'll even share a burrow with another species, probably the one who did the work. But they don't bother the animals, and they're usually in different parts of the tunnel. The skunks' improvements include a little decorating with shredded leaves or grass, and if it's really cold, they may even plug up the entrance.

During cold weather, they stay in these warm spots, slow their metabolism, and go into a deep sleep, but not a true hibernation. During warm spells, they may come out of their dens and forage.

In late winter, males venture forth seeking a mate. (Skunks are ready to mate when they're about a year old.) If a male skunk denned with one or several females, he may not need to look very far. A pair may breed and stay together until the female is ready to give birth, or the male may move on immediately. At any rate, he doesn't stay around to help raise the kits.

The female gives birth to five or six small babies after about two months. The half-ounce kits are blind and deaf at birth, but already have the characteristic black and white skunk marking on their pink skin. After about two weeks, they have grown a nice fur coat, and when they are three weeks old, they can hear and see. They weigh twelve ounces after a month of nursing, but they can barely walk and don't leave the den.

When they're about seven weeks old, and weigh about a pound and a half, they venture outside with their mother. Skunks are thorough in their explorations for food, and they inspect every possible source. The young learn to sniff every hole and to look under every rock in the small home range. They are weaned soon after they begin hunting with their mother but stay with her throughout the summer. In the cold North, the young skunks may den with the mother for the winter, but in the South, they part ways in the early fall.

Skunks live two or three years in the wild. In most areas, skunks don't have any natural enemies other than dogs. And, as fur-bearing animals, skunks are legally protected from man, outside of trapping season.

## Skunk Encounters

Deodorized skunks are coveted as pets. They're playful, intelligent (skunks know their names), and amusing. Many localities, however, outlaw pet skunks. Besides, unless you can convince your neighbors that the skunk is 100 percent aroma-free, you probably aren't going to be too popular.

As a result, most people encounter skunks in the outdoors, including, increasingly, around their yards and homes.

Skunks adapt to changes in their environment, including development. When their natural environment gets invaded, they move into residential neighborhoods in search of food and shelter. For example, in Reading, Pennsylvania, thirty-three skunks were found living in a single city block. In Chicago, a family found a skunk in a heating duct in their house. In Nelson, British Columbia, skunks had taken up residence under the plywood floor of a school and wouldn't allow music students to practice. Every time the students tuned up, the skunks showed their displeasure. One young flautist complained that it was "really hard to concentrate, especially when you have to take a big breath."

In California, meanwhile, a homeowner found fourteen laid-back skunks living (and breeding) next to his Jacuzzi. And, Marin County had such a skunk problem that the local humane society offered a seminar entitled, "Skunks: Whose House is It?"

Skunks pose four problems for homeowners. First, they spray.

While most people know enough to keep away from skunks, they're not always successful. And cats and dogs don't necessarily even know to stay away. Second, skunks can spread rabies through their bites. Third, they dig up lawns, looking for insect larvae. Fourth, they enjoy rummaging through garbage left outside (or partaking of vegetable gardens).

A lot of people are frightened at the prospect of being sprayed, and with good reason. The smell from a skunk's spray can reach up to a mile, and up close, say in a house, it can make folks feel nauseated. Besides the smell, the liquid causes eye inflammation.

There are some precautions you can take to avoid being sprayed. The first thing is to be alert if you discover—say, during a walk in the woods—that you're in skunk territory. Like backpackers who've been in the woods for a while, skunks are often smelled before they are seen. If you don't smell them, you might hear them. While they're basically silent creatures, you might hear their grunting when they feed. When agitated, the skunk may scold, growl, or snarl.

If you come across a skunk, don't advance. There's a chance it'll try to get away. Be even more wary with baby skunks, which are more likely to spray than adults. So don't walk up to a tiny skunk and say, "Oh, how cute, a baby."

What can you do if you get sprayed by a skunk? First, don't plan on a big date for that weekend. A bath in tomato juice is one prescribed course—a very expensive solution, but it will reduce the odor. After your tomato juice bath, you might consider taking a shower so you don't smell like tomatoes.

Rubbing yourself with diluted vinegar is another prescription. The most effective treatment is a liquid form of neuthroleum-alpha, available from hospital-supply shops. Wash with the solution to get rid of the skunk smell. Also, spray to the eyes may cause temporary blindness; rinse them with plenty of water. Sight should return in ten or fifteen minutes.

As for your clothes, they may well be a lost cause. But if you want to try to salvage them, you can wash them in laundry detergent and household ammonia or bleach. It helps to let them air out as much as possible. Trapper Janice Henke of Glenn Falls, New York, says that when she gets sprayed by a skunk, she takes off "every stitch," including her sneakers, and hangs everything in a tree. "I leave my clothes outside for a week or so." The utility of this remedy might depend on how close your nearest neighbors are and how much you like them.

Dogs and skunks are another issue. Dogs like to chase smaller animals, and there's nothing you can do to teach a dog that skunks are meant to be avoided. At least until it's too late. Then, when the inevitable happens, a smelly dog may only be the start of your problems, as the Minneapolis owners of Ava discovered.

One morning, Ava went out just before sunrise for a little run around the yard and found a skunk to chase. As you know, a skunk's aim is nearly perfect every time. While Ava's problems couldn't have gotten worse after she was hit, her owners' problems had just begun. Ava decided that to get the smell off her fur, she'd rub it off. So, naturally, she ran into the house and rubbed ferociously on every piece of furniture in the house.

If your dog is skunked, give it a bath with tomato juice, diluted vinegar, or neuthroleum-alpha. Vermont resident Pam Dennis used peppermint mouthwash to cleanse her dog of the skunk spray. You also might try a fresh-scented perfume to mask the odor until it fades.

People (and to the degree possible, their pets) also should avoid being bitten by a skunk. More and more skunks are becoming infected with rabies, which is fatal to humans if not treated. In some areas (most notably in the eastern United States) the number of rabid skunks exceeds the number of rabid raccoons. The increase in rabies among skunks has been attributed by some experts to the decrease in interest in skunk furs among clothing buyers: The fewer furs bought, the fewer skunks trapped, the greater the increase in the skunk population and the greater opportunity for rabies to spread. However, many biologists don't believe that trapping dampens the incidence of rabies among the population.

Rabid skunks may act aggressively, or not. They have been known to attack dogs and cats during the daytime. Perhaps one sure way to identify a rabid skunk is that a skunk with rabies won't spray (if that's any consolation)—all they want to do is bite.

A number of cities have started programs to vaccinate skunks against rabies. The vaccine is 100 percent effective on laboratory animals, but less than 80 percent effective against wild skunks because of the effect of weather and temperature on the skunk's metabolism. But the vaccine's effectiveness goes beyond these biological statistics. Because skunks are territorial, one family of inoculated skunks will fend off other, nonvaccinated skunks.

Skunks' effect on your lawn is possibly irritating, but less of a worry. Because skunks seek out grubs (insect larvae) that are relatively close to the surface, skunks don't make the kind of far-reaching holes that gophers do. They only leave a series of small, two-inch-deep holes in your lawn. Of course that's no consolation to anyone with a newly sodded lawn. But if you can bear it, nature automatically repairs skunk holes during the wet spring months. Indeed, some homeowners find that skunk holes are more desirable than the insects they eat.

But then some homeowners don't want either grub worms or skunks. In that case, one strategy involves eliminating the skunks' food source. *Bacillus popillae*, milky spore disease, will get rid of the grubs and the feeding skunks. (It's described in detail in the chapter on insects. The parasitic nematode strain Hh, also described in that chapter, will kill the grubs more quickly.)

Another way to keep skunks (and raccoons) out of your yard is not to fertilize your lawn. Fertilizers make grass lush, which attracts grubs. Cultivate a slightly less-lush lawn and the grubs, skunks, and raccoons will dine elsewhere.

At the same time that you're attacking your local grub population, you need to thwart the skunks' access to other food sources. Don't feed your pets outdoors, and secure your garbage: Put food in plastic garbage bags, place the bags inside cans, then cover garbage cans securely.

To prevent skunk problems, you also need to remove potential shelter areas. Pile fire wood on pallets that don't rest on the ground. Remove brush and piles of leaves that skunks might use for denning. They may be coming around to feed on mice in the yard, so get a mouse problem under control. Put wire mesh in front of drain openings, and use wire mesh, metal sheeting, or concrete to block crawl spaces under porches and buildings and any other holes that skunks might use to enter a house. They won't chew through wood. Skunks don't climb, so don't worry about them getting to the upper reaches of the house.

If skunks are already denning somewhere near your house, you'll probably know it. While they hardly ever spray their dens, you'll probably smell their occasional sprayings of enemies.

For dens, they like junky areas, outbuildings, garages, or holes under your house or in your yard. What you will need to do is block off the entrances to the den—without trapping the animal in the structure. (It's inhumane to the animal and also unsafe for you.) Be careful between May and August when the mother skunk is likely to have young in her den. Try to devise a one-way door so the animal can get out but can not get back in.

When the den is under your house, first seal off all possible entrances to the den but one. If you're sure there are no animals in residence, then block that hole as well.

To find out when the animal has left the den, sprinkle flour on the ground in front of the entrance, extending it about two feet in all directions. Examine it periodically after dark for evidence that the skunk has left to feed. After the den is empty, close the hole. If you don't find tracks, repeat the process for a few days.

If you don't want to do midnight repairs, you could build a one-way door that lets the skunks out but not in. It's easy enough to attach a hinged door that swings only outward, but make sure the skunk doesn't get back in. Skunks aren't dexterous like raccoons, so they'll have a hard time lifting the door. Again, do not use this stratagem in the late spring or early summer, especially during May to August, because baby skunks may be trapped inside.

If the skunks live in a hole, culvert, or similar area, an alternate approach is to try flooding their space with water. It will probably drive them out, though the skunks will announce their displeasure while leaving. Also, if flooding, be careful there are no babies who would be drowned. Mothballs or ammonia-soaked cloth will also drive them out of a den or keep them from visiting an enclosed area. Be sure not to try to drive skunks from a den when they have young, between May and August.

If a skunk wanders into your home or other building, it's best to let it leave on its own. Open all the windows and doors and leave yourself.

Peggy Pachal, a canine behavior consultant with the American Dog

Owners Association once had a skunk enter her home while she was working outdoors. Her infant lay sleeping on the floor of the house. "I was afraid the baby would start crying and startle the skunk." Thinking quickly, she opened a can of dog food and put it on the porch to lure the skunk outside. It worked, and she entered the house from another door and closed the door behind the skunk.

Occasionally skunks get trapped in window wells or other pits. You can carefully lower a board into the pit so they can climb out, and you can make the escape easy by taking the time to attach footholds of nails or cleats to the board at six-inch intervals. Approach the skunk slowly, talking softly—don't do anything suddenly. Often skunks won't spray when treated gently.

Alan and Mary Devoe relate a charming skunk tale in their book *Our Animal Neighbors*. On a walk one spring evening, they came across a skunk who had his head trapped in a tin can. Talking softly to the skunk, Alan knelt down beside it and actually petted its back. The can was tightly fitted to the skunk's head, and Devoe had to grasp the skunk and firmly pull the can away, working it back and forth, and twisting it around. Finally the can came free and Devoe slowly stood up. "The skunk sniffed at our feet, brushed his body and tail back and forth across our legs as a happy cat will do. . . . He sniffed a time or two at the snow-filled air, turned from us, and went rocking off in his crooked tracked way," they wrote.

If you have a skunk problem, you might also ask your local wildlife officials to trap and remove the animal. As with many wild animals, it's not a good idea to try and trap a skunk yourself. Many are rabid, and skunks often manage to bite the person who trapped it during transport and handling.

If you decide to capture a skunk on your own, follow the directions on caging the skunk to the letter. Think ahead to the release and attach a string or wire so you can open the door from quite a distance. Also cover the trap with canvas, leaving only the door exposed. Then when you get the trap, the skunk will feel more secure. Failing that, when the skunk is trapped, approach the cage slowly and quietly and then gently cover the trap with a canvas so the animal stays calm. It will feel secure in the dark environment. Carry the trap gently. Transport it ten or more miles away to suitable habitat (I think a truck is the best method for transportation—the canvas is no guarantee that the skunk won't spray). Good bait for a live trap includes fish-flavored cat food or bread with peanut butter or sardines.

There is a way to keep a trapped skunk from spraying. This method isn't 100-percent effective, but if you're a gutsy type, and not planning on going out for the evening with people, it might be worth a try. Once you have the skunk trapped, slowly approach the cage, holding a sheet of plastic in front of you. This way, the skunk doesn't know that a person is advancing. The plastic may also protect you from a direct hit.

Then spread the plastic over the cage and turn the cage so that it's vertical and the skunk slides to the bottom. Then use a plunger-shaped object to push the skunk into a corner. Once squeezed in place, the skunk won't spray. (This technique is used to capture and vaccinate skunks.)

Finally, a word on a way not to get rid of a skunk: shooting it at close range. This is something that many people try only once. Upon being shot, skunks spray.

## ✤ HINTS FOR SKUNKS

1. Block access to crawl spaces under the home to prevent skunks from entering.

2. Clean up garbage and eliminate access to pet food.

3. If you encounter a skunk, act calmly and move slowly to get away from it.

4. If sprayed try tomato juice, neuthroleum-alpha, diluted vinegar, and plain soap and water for your body; detergent, ammonia, bleach, and air for your clothes; tomato juice, neuthroleum-alpha, diluted vinegar, mint mouthwash, aftershave, and soap and water for your dog.

# SQUIRRELS

## TREE SQUIRRELS

From Seattle to Sarasota, Long Beach to Long Island, residents know the tree squirrel and they love or hate it; everyone has an opinion. In San Francisco an animal rights group, In Defense of Animals, sued the U.S. Navy to prevent a squirrel control program that called for the use of poisoned oats. The program stopped. In the outskirts of Washington, D.C., a farmer who accused squirrels of eating ten thousand tomatoes set his cat loose to capture, judge, and punish the rodents, and then wrote about it in the *Washington Post.* Reader response was heavily in favor of the squirrels. They're tough to beat, even in a public opinion poll.

The tree squirrel, named before anyone discovered their predilection for homes and bird feeders, were so named as to differentiate them from the ground squirrels, cute animals, but with none of the cunning and agility of their arboreal kin—although both may attack your garden. You know a tree squirrel—a fox, eastern gray, western gray, tassel-eared, or pine squirrel—from the scientific order Rodentia. Only the flying squirrel is nocturnal, the others keep regular hours so we can appreciate their antics.

Tree squirrels range virtually from coast to coast. The largest are the fox squirrels, eighteen- to twenty-seven inches long and weighing up to two-and-a-quarter pounds. The smallest are the pine and flying squirrels, eight- to fifteen-inches long and weighing up to twelve ounces. The others weigh in at about one-and-a-half pounds. Approximately half the squirrel's length is its bushy, flattened tail, which is usually held in an S-shaped curve over the body while in a sitting posture. The hind legs are larger and stronger than the front and are used when leaping from tree to tree. The front feet are adapted to holding nuts while sitting in an upright position. Squirrels come in colors ranging from black to gray to red.

The squirrel is quick and nimble. It can run, climb, and jump among branches and twigs of the loftiest trees, employing risky arboreal acrobatics. People who tend to hyperbole say that before this country was settled, a squirrel could travel treetop to treetop from the Atlantic to the Mississippi without once touching the ground. When startled on the ground, a tree squirrel usually scrambles up the nearest tree, traveling swiftly from tree to tree, seldom missing a foothold. When a squirrel exceeds its ability to jump, it may drop from the tree to the ground and scurry up another tree, apparently none the worse for any injury. Squirrels travel in lanes in trees, called "travel lanes," which they mark by scent. Squirrels will travel from pole to pole along electrical and telephone wires, occasionally on trees and houses, for city blocks without setting foot on the ground.

Squirrels have home ranges, and the size depends on the availability of food. If the population is especially high, gray squirrels have been known to migrate in search of new habitat. Typically, the home range varies from one to seven acres.

Squirrels make their intentions known with an extended vocabulary of slow warning barks, scolding, teasing, and assorted playful "chucks." When alarmed, or when another squirrel is approaching a food store, jerky tail-flickering usually accompanies these calls.

Before bird feeders, all squirrels subsisted on what they found in their habitats. Squirrels depend on mast crops such as acorns, hickory nuts, and beech nuts for food. In many parts of the country, squirrels busy themselves burying individual nuts during the late summer and early fall. Sometimes caches of nuts are stored in buildings or hollow trees, but the nuts are usually stored singly. Throughout the winter they dig up these buried or cached nuts for food. Experiments have demonstrated that gray squirrels use their highly developed sense of smell to find cached nuts. When the mast crop is poor, the squirrel is hard-pressed to obtain sufficient food. Food shortages, combined with severe winter conditions, may reduce populations and result in poor reproduction. Mature squirrels can consume up to two pounds of nuts per week. They may travel extensively in the fall to locate food if supplies in the home territory are low. Since nearly all forest wildlife species use mast, there is a direct competition for food between squirrels and other animals such as ruffled grouse, deer, black bear, chipmunks, white-footed mice, blue jays, flying squirrels, turkeys, and others. This competition points out the relationships of the forest-wildlife community.

When spring thaws occur, gray squirrels utilize new foods as they become available. Swelling buds and flowers of red and sugar maple are eaten in April. A little later, the winged fruits of red maple and American elm may be consumed. Summer foods consist of berries, mushrooms, apples, corn, and other grains. The squirrel may occasionally eat birds' eggs and chew on bones and deer antlers for calcium, phosphorus, and other necessary trace minerals. Flying squirrels share tastes with their cousins, but they eat more meat—bird eggs, birdlings, and insects. The pine squirrel eats from the same menu, but, as its name implies, prefers coniferous forests where pines provide most of its food. All squirrels will cache nuts for later use.

Squirrels are active year-round, although they may stay in the home den for several days during storms or severe winter weather. In winter, several unrelated squirrels will share the same nest. Periods of greatest activity are at dawn and in later afternoon. Although wind discourages movement, squirrels will feed in rain or snow if the wind is not strong. Squirrels are somewhat gregarious, tolerating each other in small groups except when food supplies are low.

The squirrel uses two kinds of nests: leaf nests and tree dens. Good tree dens are permanent quarters, while leaf nests serve as temporary homes during the summer. The squirrel prefers cavities in mature living trees for its winter den. The den opening is frequently formed where a branch has fallen off or a feeding woodpecker has drilled a hole into the tree. Weather, rot, and insects begin to hollow out a cavity at these locations. The gray squirrel must periodically gnaw back new callous bark tissues to keep the den entrance from sealing over. Den openings are usually about three inches in diameter.

Old hollow trees with broken tips, cracks, and many openings do not make good tree dens, but do provide hiding places for the gray squirrel. On occasion, squirrels may den in barns, garages, or attics.

During the summer months, adult squirrels frequently build and occupy leaf nests, which are usually built in the top fork of a tree or in the crotch of a limb near the trunk. Most nests are generally globular in shape, but may vary depending on the supporting base. A single entrance usually faces the main tree trunk or the nearest limb that provides access to the nest. Three or four structural parts make up the nest; the base and support are made of twigs from the nest tree, the floor of the inner chamber is made of a layer of compact humus and organic debris mixed with twigs, and the outer shell is composed of leaves and twigs. In many cases, an inner shell of woven bark, grass, and leaves may be constructed to provide warmth and added protection. The outside diameters of these nests range from fourteen to sixteen inches, and they weigh from six to seven pounds. Leaf nests may serve as temporary quarters near seasonal food supplies. They are also cooler and free of the parasites that accumulate in winter dens.

Squirrels mate once a year, except for the fox and gray squirrels, which mate between December and January and again in the early summer. One reason squirrels are so difficult to breed in captivity is that a female's fertility seems to be boosted after being chased helter-skelter through the trees by one, two, or more males. After a forty-four-day gestation period, the female gives birth to three bald, blind, earless offspring. The young are dependent on the female parent for nearly six weeks; their eyes do not open until around the thirty-seventh day. Young from the first litter venture from the nest or den in early May, while members from the second litter become active in early August. They climb about the den tree for a few days before actually descending to the ground for the first time. Spring litters are usually born in a tree den, while summer litters may be born in leaf nests. Many times, the number of females producing a second litter depends on the availability of food. Squirrels (especially males) do not keep mates for life, and may migrate between litters. As a consequence, genetic inbreeding among squirrels is rare.

Natural predators include man, hawks, owls, fox, bobcats, and raccoons. House cats also prey on young squirrels. Road kills are usually high in the fall when squirrels are prone to migrate longer distances in search of food. The average life span of gray squirrels is eighteen months, but they have been known to live to eight years. Mortality is high in the wild, and about half of the squirrel population dies each year. While squirrels can live up to twelve years in captivity, an old wild squirrel is about four years old. Although lots of animals like to eat squirrels, predation doesn't have much of an effect on the population. Plenty of people hunt squirrels, even folks without a garden grudge who just like squirrel meat. The drawback?

"They look like skinned rats," says Barbara Meadows, wife of a dedicated West Virginia hunter who lovingly filled her sink with the rodents. He has to cook his own.

Squirrels are notoriously tough to outwit. I know, I wrote the book on them: *Outwitting Squirrels: 101 Cunning Stratagems to Reduce Dramatically the Egregious Misappropriation of Seed from Your Bird Feeder by Squirrels* (Chicago Review Press, 1988). It's a terrific book, but since writing it, I've discovered some more tricks on keeping squirrels out of your home and bird feeder.

One technique I discovered inadvertently while being interviewed for a television show's special segment on squirrels. (I call this technique the grab-anything technique.) The interviewer and I were standing on the back deck of my house. After talking about squirrels, the interviewer and I started walking around the yard so that he could photograph squirrels in action. During this time we had thrown nuts all over the deck and yard to attract squirrels. As we were looking around the yard I noticed a squirrel walk up to the back door, but I didn't see the squirrel leave the vicinity of the door. It was at that moment that I realized that the squirrel had gone inside our kitchen through the open door. I raced up the stairs (the deck was on the second floor) screaming for my wife (who was at that moment tending to our screaming baby, and so she noticed nothing). The cameraman raced up behind me, apparently not intending to help, but hoping to film the episode for television. He did. Luckily I managed to shoo the squirrel back out—a trick that was more a matter of luck than skill.

There are two schools of thought about squirrels taking up house-keeping in your home. One is ruthless, believing a squirrel will always remember what a lovely home you have and attempt to return. The other is patient, believing the squirrel will eventually give up and live in a tree.

John Adcock, a pest control professional says, "You can't chase squirrels out of your house. Squirrels have to be trapped. They'll never leave!" He bases his advice on experience. When I spoke with him he had just removed a faccia board from a home in Chevy Chase, Maryland. It had thirty-seven holes in it, and that's just one board. There were four faccia boards on this home which represented fifteen years of exclusion where the people had dealt with squirrels by simply chasing them out and closing up the hole. "It doesn't work; they just chew another hole in the house. They have a vested interest in the house, due to the fact that they have stored food there. This is where they're planning on raising their young," Adcock sums up. All without having to get a mortgage!

And some times are worse than others for plugging up squirrel holes. Sometimes folks patch up a squirrel access hole only to find a crazy squirrel that's chewing holes in the house. They throw stones and yell, and the squirrel keeps it up. Imagine if someone blocked off the door to your nursery—you'd go through anything to get to the baby. Well, a squirrel mother is the same way. Get that access opened up, let her get to her babies, and she'll be all right.

People do things after the fact. The time to screen is not during the middle of a squirrel infestation. And people declare victory too soon. In the summer, attics can reach temperatures of 120 or 130 degrees Fahrenheit, and no squirrel in its right mind is going to go up there. When it gets hot like that, people think the squirrels have moved on and that there is no need for repair. In reality the squirrel was ready to leave that area anyway. Come next November or September, however, they're back there hammering on the house, chewing another hole. But now the guy who had one hole in his house has two, due to the fact that he never took care of the problem. The first priority is to get rid of the squirrels that have the interest in your property. They consider it their home. You're making it your storage space, but it's their home.

Becca Schad of Wildlife Matters concurs. "You get whole generations of squirrels who were born in attics and think that's where they're supposed to live." A tree nest looks pretty insecure; the wind could blow, it could fall out. But an attic looks pretty good, warm, and dry; it's perfect. Squirrels are probably the most destructive animal in an attic. They gnaw everything: insulation off wires and boxes, stored clothes, and your tax records from 1987.

Schad represents the patient school. Like Adcock, she would never attempt to oust a mother squirrel, but once the babies are grown, she becomes ruthless, too; that is, she blocks up every hole imaginable. And she cuts back branches that allow squirrels to get to the roof or attic.

John Hadidian of the National Park Service says, "Squirrels may be scared away with strips of mylar tape on the fence or a rag soaked in ammonia."

Outdoors, squirrels are just as tenacious. "What happens often is you get populations that have a tradition. You get tomato-eating squirrels, and the mother teaches the young that this is a food source. Three blocks away, the squirrels don't eat tomatoes. You have to teach the squirrels in your neighborhood that tomatoes aren't food. But you ought to be able to do that and you ought to count on taking two years to do it."

Hadidian pointed out that mothballs, often used indiscriminately outdoors, are systemic compounds that are absorbed into plants. Something like dried-blood fertilizer may do a better job outdoors, and it will help the plants. However, used indoors in enclosed spaces, mothballs or flakes may encourage animals to leave long enough for you to close off a point of entry.

John Adcock has a different view: "Mothballs are camphor. Camphor is actually flammable. To put camphor in an area that gains temperatures of 120–140 degrees Fahrenheit is ludicrous, as far as spontaneous combustion is concerned." He wonders how many mothballs are necessary in a two-thousand-square-foot house with nine hundred square feet under the roof. "What do you do? Put a mothball every six inches? If you do, they just kick them out of the way. You'd have to fill that whole attic full of camphor to make it work."

Lowell Robertson of Sonic Technology, a California manufacturer of Pest Chaser ultrasonic devices, got rid of the squirrels in his attic.

"The devices seem to work really well on squirrels. I'll tell you straight off that we have not done specific testing on squirrels, but I know from my own experience that it got squirrels out of my attic. Squirrels can apparently perceive a portion of the ultrasonic band that we're broadcasting."

There is a lot of debate over whether ultrasonic devices work to control rodents. The manufacturers and retailers say they do; almost everybody else says they don't. But does it really matter? These sonic devices are like elephant whistles: You've heard the story about the New Yorker who blows on the odd-sounding whistle. His friend asks, "What is that whistle for?" The whistle-blower replies, "It's to keep elephants away." His friend, incredulous, asserts, "No way can that work." To which the whistler says, "But do you see any elephants around?"

"Squirrels can eat right through a two-by-four. When you exclude them from a nesting area they like, they just chew another hole. I literally pounded up twenty square feet of hardware cloth at all the points where I had evidence of them getting in. And they just chewed another hole. That's when I finally put one of my own products in my attic. And that cured the problem. Many people have said that they have had the same experience with the product," says Robertson.

Diane White, a writer for the *Boston Globe*, and her cat were helpless against a female squirrel. The squirrel entered the home through the cat's door and helped herself to English muffins, croissants, and chocolates. While her husband suggested a BB gun solution, she advocated the do-nothing approach, not wanting to keep her cat from the litter box on the other side of the swinging pet door.

Whether they're in your attic or in the yard, the only solution may be to trap and relocate an especially pesky squirrel. Garon Fyffe, director of ABC Humane Wildlife Rescue and Relocation near Chicago, relocates about eight hundred squirrels a year, moving the varmints from where they're unwelcome, and taking them to spots where they're wanted. He live-traps the animals and often releases them in wildlife preserves, where he has a permit to release the animals. He's careful to make sure the preserves don't abut neighborhoods where the squirrels will become problematic to someone else.

It's easy enough to capture the little critters and take them for a little country ride to an oak tree stand, but how do you know they won't just return? Squirrels migrate to some extent; but they're not homing pigeons, so they probably won't return to your home.

## At the Bird Feeder

A Canadian naturalist and animal rights activist, Barry Kent MacKay, decided he had met a new squirrel species, the *Sciurus carolinensis hellensis*, the gray squirrel from hell. They attacked his flowers, his bird seed, and his bird feeders, "chewing through plastic as if it were compressed peanut butter." He tried everything: squirrel baffles, squirrel-proof feeders, greasing the bird feeder's pole, moving the bird feeder pole, installing squirrel deterrents on the feeding platform, and cutting back branches on trees all around the feeder until "the yard looked like a war zone." Eventually he drove the squirrels to the ground, where they are subsisting on the dropped seeds from the birds, but he knows it's only a matter of time until the demonic animals figure out a new access to the seed supply.

In other parts of North America, there are less tenacious squirrels. Often, greasing the bird feeder's pole works, especially if you mix Vaseline and cayenne pepper or curry powder. Other greasy concoctions containing axle grease or solid vegetable shortening are off limits: Axle grease is lethal, and the shortening often spoils and sickens the squirrel.

Generally, squirrel-proof bird feeders are effective. There are feeders that only open when a selected bird alights on the perch. You set the dial to the bird you want to feed, and a system of counterweights opens only to a bird of that weight. Since birds of a species weigh more or less the same, there will be no anorexic starlings eating from the feeder you have set for a finch. Other feeders use wire to keep the squirrel away from the feeding tube—pick one that won't allow the squirrel to get its teeth against the plastic or it will eat the feeder. Another feeder that's virtually squirrel-proof is the MH Industries GSP (Guaranteed Squirrel Proof) Feeder. Only animals that can fly and small helicopters can get inside the globe that opens at the bottom. The feeders that operate by closing a door in front whenever something heavy lands are effective, too. These go by names like the Absolutely Squirrel-Proof Feeder, Squirrel's Defeat, the Foiler, and the Steel Squirrel-Proof Feeder.

Try hanging the feeder on a small chain the squirrel can't chew, and install a baffle above the feeder. Some squirrels aren't especially brainy. Robert Dewey of Washington, D.C., baffled squirrels with no more than a metal coffee can lid strung onto the hanging wire. (No, it doesn't have sharp edges.) But it was enough to convince the squirrels to feed on the ground spills with the mourning doves.

## Distractions

If you can't beat 'em, feed 'em. That's what some people eventually decide. Give the squirrels a feeding station of their own, and maybe they'll leave the birds alone.

A "three-ring circus" or "squirrel-a-twirl" is a three-armed windmill with an ear of corn attached to the end of each arm. Mount the contraption on a tree or any flat surface. The device rotates as the squirrels try to get the corn—and they will try. If that's not silly enough, you can get a table and chair set just for Squirrel Nutkin. An ear of corn stands upright on a spike on the table, and the squirrel sits in the chair as it feeds. Double your fun with a setting for two.

## Flat Squirrels

I think we'll all agree that a city is a dangerous enough place, but despite all the predators, there are few for the squirrel. In fact, the car is probably the main predator of the squirrel in cities. It's not that squirrels aren't fast and wily, they just have rotten depth perception and close-up vision. Add to that the confusion and noise of traffic, and you have a plausible explanation for the squirrel carnage on our roadways.

## Squirrel Trivia

And last, failing all the above solutions, here is some interesting squirrel trivia that may—or may not—help you get rid of troublesome squirrels. Remember, the more you know about squirrels the more successful your strategies will be.

• In 1834 there was a competition between two towns in Indiana to see who could shoot the most squirrels in one day. The winner bagged nine hundred; the runner up, seven hundred.

• In the original Norman-French version of the fairy tale, Cinderella, her slippers were made of squirrel fur.

• Gray squirrels were introduced into Great Britain in 1876. In 1937 the British prohibited the importation of squirrels without a license. (They're pushing out the British native, the red squirrel.)

• In Great Britain you need a license to keep squirrels as pets.

• In the early 1900s because of hunting and deforestation, but mostly because of the latter, there was some concern that gray squirrels could become extinct.

• In Olney, Illinois, motorists who hit a squirrel are fined $25.

• After Hurricane Gloria swept the eastern United States in 1985, more squirrels were spotted on lawns and sidewalks than ever before in Long Island. These squirrels were called "displaced, neurotic squirrels" by one *New York Times* reporter.

• The squirrel is the state mammal in North Carolina.

• Washington, D.C.'s first wild black squirrels were said to be escapees from the National Zoo. Twenty-eight black squirrels from Ontario were brought to the zoo between 1902 and 1906.

• Squirrels must eat most of the time that they are active.

• Squirrels' digestive tracts are over seven feet long.

• Ground squirrels consume twenty-four ounces of nuts a week. They can eat three ounces at a sitting.

• The average number of squirrel road kills is one per every ten miles.

• Female adult squirrels live longer than males.

• When interested in mating, male squirrels follow females for five days.

## ❖ SQUELCHING SQUIRRELS

1. Install squirrel baffles around birdfeeders.

2. Block squirrels from entering your home; if they obsess about living with you, provide a squirrel nesting box on a nearby tree.

3. Keep tree branches trimmed back so they don't provide easy access to your attic.

4. Get a squirrel feeder.

# ENDNOTE

The really frightening thing is that nearly a year later the trap is still unsprung. I know I heard and saw something fuzzy that evening. And it saw me.

There are two possibilities. First, the chipmunk has escaped to the outside. That's the possibility I'm counting on.

The other possibility is that the chipmunk is still here. And we won't find out until we renovate again.

# AFTERWORD

Half a dozen years have passed since I first started writing *Outwitting Critters*. During these years, I have had many more encounters with various animals than ever before—without leaving my home or venturing far from it. There is now a herd of deer a mile or so away in Rock Creek Park, a long winding woods that runs the length of Washington, D.C. A few weeks ago, during one of Washington's hot, humid summer days, a raccoon decided to take an afternoon snooze on a shaded windowsill. The kids loved the show—my wife and I could only think "rabies." And a family—more like an entire neighborhood—of woodpeckers lives very, very close by.

We live in a fairly urban setting, yet we are visited by opossums, raccoons, squirrels, mice, animals that look like mice but are much bigger, deer, ravens, bugs of all varieties, and even the occasional pigeon. As urban and suburban sprawl persist, nature likes to remind us that it's still the boss by introducing its friends to us every now and then. Well, "introducing" may be too mellow a word: Nature likes to barge in every now and then to remind us that there is more to the planet than just people and our things.

Much of what I learned while writing the first edition of *Outwitting Critters* is still very true: No matter how hard we try to keep critters out of our houses, gardens, birdfeeders, swimming pools, cars (a friend and I once returned from a twelve-day expedition in the White Cloud Mountains of Idaho only to discover that a family of mice had broken into his car with the skill of the most savvy New York car thief), clothes, backyards, and other places where we like to have serenity and sanctuary, they always are a step ahead of us.

Critters haven't grown any smarter, but over the years we have moved our homes closer to places where Nature rules, and where we are just visitors. How can you blame a deer for snacking on turnips in the deer's backyard? Or an alligator for taking a swim in the nearby swimming pool that to him feels just like the swamp in which he lives? Or a squirrel for deciding that inside your house is much warmer than inside the hollow of a tree? From the perspective of many critters, *we* have invited *them* into our yards and homes (not to mention cars and shoes).

Fortunately, the techniques that work best are also the ones that don't injure animals. Harming or killing an animal is not only inhumane, it is certain to make you feel lousy. Never mind that when you hurt an animal you're likely to inspire the wrath of all of that species (okay, so that's only a nightmare I used to have as a child), but in the long term, you often have to deal with many animals, not a single critter. Killing a lone gopher isn't going to help you much, where there are several families of gophers turning your yard into a shabby-looking golf course.

Besides, it's fun and interesting to outwit critters. We're supposed to be the most intelligent beings on the planet, and while many of us can't prove that point by our chess-playing skills or ability to swiftly toilet train a toddler, we can display our mental prowess by outwitting critters. How to outwit them is what this book is about. The short answer is you have to learn as much as you can about the creatures you intend to outmanuever; you need to get down on your hands and knees, crawl about and look at the world from the perspective of an animal. (If you are trying to outwit crows and want to explore the world from a crow's point-of-view, proceed at your own risk.) The long answer is: Read this book. These pages contain a great deal of experience and wisdom from people who have succeeded, and failed, in their efforts to outwit critters.

Good luck.

If you come across any new critter techniques, please write me at:

Adler & Robin Books, Inc.
3000 Connecticut Avenue, N.W.
Washington, DC 20008

—Bill Adler, Jr.
Spring 1997

# RESOURCES

Thanks to the Great Plains Agricultural Council Wildlife Resources Committee and the Cooperative Extension Service at the University of Nebraska-Lincoln for much of the information in this section.

## GROUPS

American Dog Owners Association
1920 Route 9
Castleton, NY 12033
518-477-8469

Bat Conservation International
P.O. Box 162603
Austin, TX 78716-2603
A wealth of information on bats, in addition to bat houses and
    plans for bat houses.

The Beaver Defenders
Unexpected Wildlife Refuge, Inc.
Box 765
Newfield, NJ 08344
609-697-3541

Friends of Beaversprite
Box 591
Little Falls, NY 13365

The Fund for Animals
850 Sligo Avenue, Suite LL2
Silver Spring, MD 20910
301-585-2591

National Institute for Urban Wildlife
10921 Trotting Ridge Way
Columbia, MD 21044
301-596-3311
A private, nonprofit, scientific, and educational organization
    dedicated to the conservation of wildlife in urban, suburban,
    and developing areas.

Wildlife 2000
Ms. Sherri Tippie
P.O. Box 6428
Denver, CO 80206
303-935-4995
Beaver and all wildlife

Zebra Mussel Information Clearinghouse
Susan Grace Moore, Librarian
State University of New York
Brockport, NY 14420
716-395-2638

# PRODUCTS

## Bats

### *Bat exclusion materials and devices*

Bay Area Bat Protection
1312 Shiloh Road
Sturgeon Bay, WI 54235

3 E Corp
401 Kennedy Boulevard
P.O. Box 177
Somerdale, NJ 08083
609-784-8200

### *Bat houses*

Bat Conservation International
P.O. Box 162603
Austin, TX 78716-2603

## Birds

### *Electronic alarms and recorded sounds*

#### ALARMS OR RECORDED DISTRESS CALLS

Applied Electronics Corporation
3003 County Line Road
Little Rock, AR 72201
501-821-3095

Av-Alarm Corp.
675-D Conger Street
Eugene, OR 97402
503-342-1271

Margo Horticultural Supplies, Ltd.
RR 6, Site 8, Box 2
Calgary, Alberta T2M 4L5
Canada
403-285-9731

Signal Broadcasting Co.
2314 Broadway Street
Denver, CO 80205
303-295-0479

### AUTOMATIC EXPLODERS

Alexander-Tagg Industries
395 Jacksonville Road
Warminster, PA 18974
215-675-7200

C. Frensch Ltd.
168 Main Street, East
Box 67
Grimsby, Ontario L3M 4G1
Canada
416-945-3817

B. M. Lawrence and Co.
233 Sansom Street
San Francisco, CA 94104
415-981-3650

Peaceful Valley Farm Supply
P.O. Box 2209
Grass Valley, CA 95945
916-272-4769

Margo Horticultural Supplies, Ltd.
RR 6, Site 8, Box 2
Calgary, Alberta T2M 4L5
Canada
403-285-9731

Pisces Industries
P.O. Box 6407
Modesto, CA 95355
209-578-5502

Reed-Joseph International Co.
P.O. Box 894
Greenville, MS 38702
601-335-5822
800-647-5554

Smith-Roles
1367 South Anna Street
Wichita, KS 67209
316-945-0295
*or*
Box 1607
Minot, ND 58702
701-852-3726

Teiso Kasei Co., Ltd.
350 South Figueroa Street, Suite 350
Los Angeles, CA 90071
213-680-4349

### OTHER SOUND-PRODUCING DEVICES

Falcon Safety Products, Inc.
1065 Briston Road
Mountainside, NJ 07092
201-233-5000

Tomko Enterprises, Inc.
Route 58, RD 2
P.O. Box 937-A
Riverhead, NY 11901
516-727-3932

## *Kites, balloons, raptor, and human effigies*

### BALLOONS ONLY

Peaceful Valley Farm Supply
P.O. Box 2209
Grass Valley, CA 95945
916-272-4769

Raven Industries
P.O. Box 1007
Sioux Falls, SD 57117
605-336-2750

WeatherMeasure Corp.
P.O. Box 41257
Sacramento, CA 95841

### BALLOONS, MYLAR TAPES, AND KITES

Atmospheric Instrumentation Research (AIR), Inc.
1880 South Flatiron Court, Suite A
Boulder, CO 80301
303-443-7187

Brookstone
127 Vose Farm Road
Peterborough, NH 03458
603-924-9541

### FLASHING OR REVOLVING LIGHTS

Bird-X
325 West Huron Street
Chicago, IL 60610
312-642-6871

R. E. Dietz Co.
225 Wilkinson Street
Syracuse, NY 13201
315-424-7400

The Huge Co.
7625 Page Boulevard
St. Louis, MO 63133
314-725-2555

Tripp-Lite Manufacturing Co.
500 North Orleands
Chicago, IL 60610
312-329-1777

### KITES

Sutton Ag Enterprises
1081 Harkins Road
Salinas, CA 93901
408-422-9693

Teiso Kasei Co., Ltd.
350 South Figueroa Street, Suite 350
Los Angeles, CA 90071
213-680-4349

Tiderider, Inc.
P.O. Box 9
Eastern and Steele Boulevards
Baldwin, NY 11510
516-223-3838

### RAPTOR EFFIGIES

W. Atlee Burpee Seed Co.
Warminster, PA 18974
215-674-4900

Bird-X
325 West Huron St.
Chicago, IL 60610
312-642-6871

The Huge Co.
7625 Page Boulevard
St. Louis, MO 63133
314-725-2555

Peaceful Valley Farm Supply
P.O. Box 2209
Grass Valley, CA 95945
916-272-4769

Plow and Hearth Catalog
301 Madison Road
P.O. Box 830
Orange, VA 22960
800-627-1712

Teiso Kasei Co., Ltd.
350 South Figueroa Street, Suite 350
Los Angeles, CA 90071
213-680-4349

### SCARECROWS

W. Atlee Burpee Seed Co.
Warminster, PA 18974
215-674-4900

## Netting

Almac Plastics, Inc.
6311 Erdman
Baltimore, MD 21205-3585
(Conwed)

Conwed Corporation
Plastics Division
P.O. Box 43237
St. Paul, MN 55164-0237
612-221-1260

Green Valley Blueberry Farm
9345 Ross Station Road
Sebastopol, CA 95472
707-887-7496

Internet, Inc.
2730 Nevada Avenue, North
Minneapolis, MN 55427
612-541-9690

Orchard Supply Co. of Sacramento
P.O. Box 956
Sacramento, CA 95804
916-446-7821

Teitzel's Rainier View Blueberry Farms
7720 East 134th Avenue
Puyallup, WA 98371
206-863-6548

Wildlife Control Technology
6408 Fig St.
Fresno, CA 93706
209-268-1200

Animal Repellents, Inc.
P.O. Box 999
Griffin, GA 30224
404-227-8222
800-241-5064
(Durex)

J. A. Cissel Co., Inc.
P.O. Box 339
Farmingdale, NJ 07727
201-938-6600
(Toprite)

Bob Ellsworth—The Complete Winemaker
1219 Main Street
St. Helena, CA 94574
707-963-9681
(Xironet)

Margo Horticultural Supplies, Ltd.
RR 6, Site 8, Box 2
Calgary, Alberta T2M 4L5
Canada 403-285-9731

## Other netting

A to Z Net Man
P.O. Box 2168
South Hackensack, NJ 07606
201-488-3888

Bird-X
325 West Huron Street
Chicago, IL 60610
312-642-6871

Blue Mountain Industries
20 Blue Mountain Road
Blue Mountain, AL 36201
205-237-9461

J. A. Cissel Co., Inc.
P.O. Box 339
Farmingdale, NJ 07727
201-938-6600

SINCO
P.O. Box 361
East Hampton, CT 06424
203-267-2545

## Porcupine wire

Nixalite of America
417 25th Street
Moline, IL 61265
309-797-8771
(Nixalite)

Shaw Steeple Jacks, Inc.
2710 Bedford Street
Johnstown, PA 15904
814-266-8008
(Cat Claw)

## Pyrotechnic devices (fireworks, shellcrackers, whistle bombs, etc.)

Marshall Hyde, Inc.
P.O. Box 497
Port Huron, MI 48060
313-982-2140

Margo Horticultural Supplies, Ltd.
RR 6, Site 8, Box 2
Calgary, Alberta T2M 4L5
Canada
403-285-9731

New Jersey Fireworks Co.
Box 118
Vineland, NJ 08630
609-692-8030

O. C. Ag Supply
1328 Allec Street
Anaheim, CA 92805

Stoneco, Inc.
P.O. Box 187
Dacono, CO 80514
303-833-2376

Sutton Ag Enterprises
1081 Harkins Road
Salinas, CA 93901
408-422-9693

Wald and Co.
208 Broadway
Kansas City, MO 64105
816-842-9299

Western Fireworks Co.
2542 13th Avenue, SE
Canby, OR 97013
503-266-7770

### *Sticky Substances*

ArChem Corp
1514 11th Street
P.O. Box 767
Portsmouth, OH 45662
614-353-1125

Bird Control International
J. T Eaton and Co.
P.O. Box 12
Macedonia, OH 44056
216-425-2377

Crown Industries
4015 Papin Street
St. Louis, MO 63110
314-533-0999
800-325-3316

J. C. Ehrlich Chemical Co.
State College Laboratories
840 William Lane
Reading, PA 19612
215-921-0641


Hub States Corp.
419 East Washington Street
Indianapolis, IN 46204
800-428-4416


The Huge Co.
7625 Page Boulevard
St. Louis, MO 63133
314-725-2555


Sun Pest Control
2945 McGee Trafficway
Kansas City, MO 64108
816-561-2174


The Tanglefoot Co.
314 Straight Avenue, SW
Grand Rapids, MI 49504
616-459-4130


Velsicol Chemical Co.
341 E. Ohio Street
Chicago, IL 60611
312-670-4500

# Bugs

*Traps, botanical pesticides, boric acid, barriers, shampoos, sticky bug traps, beneficial insects*

The Alsto Company
P.O. Box 1267
Galesburg, IL 61401
800-447-0048
(lawn sandals)

Brookstone
127 Vose Farm Road
Peterborough, NH 03458
603-924-9541
(dryer vent sealer)

Gardeners Supply Company
128 Intervale Road
Burlington, VT 05401
800-548-4784

Gardens Alive!
Natural Gardening Center
Highway 48
P.O. Box 149
Sunman, IN 47041
812-623-3800

Melinger Nursery
2310 West South Range Road
North Lima, OH 44452
216-549-9861
(aerator sandals)

Necessary Trading Company
8311 Salem Avenue
New Castle, VA 24127
800-447-5354

Ringer Corporation
9959 Valley View Road
Eden Prairie, MN 55344
800-423-7544


Rincon-Vitova Insectaries
P.O. Box 95
Oak View, CA 93022
800-248-2847
In CA: 800-643-5407


Seventh Generation
Colchester, VT
802-655-3116
(dryer vent)


Unique Insect Control
5504 Sperry Drive
Citrus Heights, CA 95621
916-961-7945


# Mammals

## *Beaver architectural solutions*


D.C.P. Consulting, Ltd.
Beaver Stop Division
3219 Coleman Road, NW
Calgary, Alberta, T2L 1G6
Canada
403-282-2506
403-282-6136

## *Browsing mammal exclusion devices*

Almac Plastics
6311 Erdman
Baltimore, MD 21205-3585
301-485-9100

Forest Protection Products
1420 North Seventh Street
P.O. Box 1557
Coos Bay, OR 97420
503-267-2622

## *Burrowing animals—vibrating stakes*

Handsome Rewards
19465 Brennan Avenue
Perris, CA 92379
714-943-2023

## *Frightening devices*

Plow and Hearth
301 Madison Road
P.O. Box 830
Orange, VA 22960
800-627-1712

Sporty's
Clermont Airport
Batavia, OH 45103-9747
800-543-8633
(Electric scat matt)

Tomko Enterprises, Inc.
Route 58, RD 2
P.O. Box 937-A
Riverhead, NY 11901
516-727-3932

## *Live traps*

BD Tru-Catch Inc.
Box 327
Dickinson, ND 58602-0327
701-225-0398

B-Kind Animal Control Equipment
Southeastern Metal Products, Inc.
1200 Foster Street, NW
P.O. Box 93038
Atlanta, GA 30377
404-351-6686

Hancock Trap Co.
Box 268
Custer, SD 57730
605-673-4128

Holdzem Trap Division
Oberlin Canteen Co.
212 Sumner Street
P.O. Box 208
Oberlin, OH 44074
216-774-3391

Ketch-All Co.
2537 University Ave.
San Diego, CA 92104
619-297-1953

Mustang Manufacturing Co.
P.O. Box 10947
Houston, TX 77018
713-682-0811

Pioneer Wildlife Traps
2909 Alberta Street, NE
Portland, OR 97211
503-249-2935

H. J. Spencer and Sons
P.O. Box 131
Gainesville, FL 32602
904-372-4018

Stendal Products, Inc.
986 East Laurel Road
Bellingham, WA 98226
206-398-2353

Sullivan's Sure-Catch Traps
Box 1241
2324 South Patterson
Valdosta, GA 31601
912-242-1677

Tomahawk Live Trap
P.O. Box 323
Tomahawk, WI 54487
715-453-3550

Woodstream Co.
Lititz, PA 17543
717-626-2125

### *Predator calls (to deter prey animals or attract predators)*

Burnham Brothers
P.O. Box 669
Marble Falls, TX 78654
800-451-4572

Circe Calls
P.O. Box 697
Goodyear, AZ 85338
602-971-7182

## *Repellents*

### AMMONIUM SOAPS OF HIGHER FATTY ACIDS

Leffingwell Division
UNIROYAL Chemical Co.
111 South Berry Street
P.O. Box 1880
Brea, CA 92621
714-529-3973

### BONE TAR OIL

J. C. Ehrlich Chemical Co.
State College Laboratories
840 William Lane
Reading, PA 19612
215-921-0641


Planttabs Co.
Box 397
Timonium, MD 21093
301-252-4620


### CAPSAICIN (HOT)

Miller Chemical and Fertilizer Corporation
Box 333
Hanover, PA 17331
717-632-8921


Seventh Generation
Colchester, VT
802-655-3116

Deer-Away
McLaughlin Gormley King Co.
712 15th Avenue, NE
Minneapolis, MN 55413
612-379-2895

## Sticky repellent glues and traps

Airchem Corp
1514 11th Street
P.O. Box 767
Portsmouth, OH 45662
614-353-1125

Animal Repellents, Inc.
P.O. Box 999
Griffin, GA 30224
404-227-8222
800-241-5064

J. T. Eaton and Co.
1393 Highland Road
Twinsburg, OH 44087
216-425-7801

Hub States Corp.
419 East Washington Street
Indianapolis, IN 46204
317-636-5255
800-428-4416

Southern Mill Creek Products
P.O. Box 1096
Tampa, FL 33601
813-626-2111

Tanglefoot Co.
314 Straight Avenue, SW
Grand Rapids, MI 49504
616-459-4130

### GLUE BOARDS

A&B Exterminators and Manufacturing, Inc.
Star Exterminating Co.
5911 Church Avenue
Brooklyn, NY 11203
212-498-1200

ABEPCO Manufacturing
1350 Collins, Unit C
Orange, CA 92667
714-538-1444

American Fluoride Corp
17 Huntington Place
New Rochelle, NY 10801
914-235-6925

Available Exterminators and Manufacturing, Inc.
P.O. Box 137
Brooklyn, NY 11236
212-375-5333

Bell Laboratories
3699 Kinsman Boulevard
Madison, WI 53704
608-241-0202

Brody Enterprises
9 Arlington Place
Fair Lawn, NJ 07410
201-794-3616

J. T. Eaton and Co.
1393 Highland Rd.
Twinsburg, OH 44087
216-425-7801

Hampton Chemical Co.
P.O. Box 1034
Riverdale, NY 10471
212-966-3081

Motomco Ltd.
29 North Fort Harrison Avenue
Clearwater, FL 33515
813-447-3417

Pest Control Supplies
1700 Liberty Street
P.O. Box 5668
Kansas City, MO 64102
316-421-4969
800-821-5689

Sherman Technology
76 Ninth Avenue
New York, NY 10011
212-242-5008

Southern Mill Creek Products
P.O. Box 1096
Tampa, FL 33601
813-626-2111

Stone Chemical Labs
467 N. Aberdeen Street
Chicago, IL 60622
312-733-6554

Tanglefoot Co.
314 Straight Avenue, SW
Grand Rapids, MI 49504
616-459-4130

Woodstream Corp.
Lititz, PA 17543
717-626-2125

## Pets

Animail Pet Care Products
P.O. Box 854
Clearfield, Utah, 84015
800-255-3723
(everything under the sun)

Audubon Workshop
1501 Paddock Drive
Northbrook, IL 60062
800-325-9464

Gardens Alive!
Natural Gardening Center
Highway 48, P.O. Box 149
Sunman, IN 47041
812-623-3800

Sporty's
Clermont Airport
Batavia, OH 45103-9747
800-543-8633

# PUBLICATIONS

*Common Sense Pest Control*
by Sheila Daar and Helga and William Olkowski

Bio Integral Resource Center
Box 7414
Berkeley, CA 94707
415-524-2567

*Safety at Home: A Guide to the Hazards of Lawn and Garden Pesticides and Safer Ways to Manage Pests*
National Coalition Against the Misuse of Pesticides
701 E Street, SE, Suite 200
Washington, D.C. 20003

# BOOKS

*The Art of Raising Puppy,* the Monks of New Skete
*Catwatching,* Desmond Morris
*Dogwatching,* Desmond Morris
*Don't Shoot the Dog,* Karen Pryor
*The Educated Cat,* Susan Fadem and George Ney
*Encyclopedia of North American Birds,* John K. Terres
*Good Dog, Bad Dog,* Mordecai Siegal and Matthew Margolis
*How to Be Your Dog's Best Friend,* the Monks of New Skete
*How to Toilet-Train Your Cat: 21 Days to a Litter-Free Home,* Paul Kunkel
*Man Meets Dog,* Konrad Lorenz
*No Bad Dogs,* Barbara Woodhouse
*Supercat,* Michael W. Fox

# MAIL-ORDER CATALOGS WITH A VARIETY OF PRODUCTS FOR PETS, BIRDS, AND MAMMALS

Audubon Workshop
1501 Paddock Drive
Northbrook, IL 60062
800-325-9464

Brookstone
127 Vose Farm Road
Peterborough, NH 03458
603-924-9541


David Kay Catalogue for Home and Garden
One Jenni Lane
Peoria, IL 61614-3198
800-535-9917


Gardens Alive!
Natural Gardening Center
Highway 48, P.O. Box 149
Sunman, IN 47041
812-623-3800


Gardener's Eden
P.O. Box 7307
San Francisco, CA 94120-7307
415-421-4242


Handsome Rewards
19465 Brennan Avenue
Perris, CA 92379
714-943-2023


Handy Helpers
P.O. Box 1267
Galesburg, IL 61401
800-447-0048


Jackson and Perkins
Medford Oregon 97501
800-292-4769

Peaceful Valley Farm Supply
P.O. Box 2209
Grass Valley, CA 95945
916-272-4769

Plow and Hearth
301 Madison Road
P.O. Box 830
Orange, VA 22960
800-627-1712

Seventh Generation
Colchester, VT 05446-1672
800-456-1171

Solutions
P.O. Box 6878
Portland, OR 97228
1-800-342-9988

Sporty's
Clermont Airport
Batavia, OH 45103-9747
800-543-8633